New National Framework

MATHEMATICS 9

WORKBOOK

Maryanne Tipler
Rachel Bates

Published in 2004 by:
Nelson Thornes Ltd
Delta Place
27 Bath Road
CHELTENHAM
GL53 7TH
United Kingdom

13 / 10 9 8 7 6 5 4 3 2

A catalogue record for this book is available from the British Library

ISBN 978 0 7487 9140 8

Illustrations by Ian West
Page make-up by Mathematical Composition Setters Ltd

Printed in China

Acknowledgements

The publishers would like to thank QCA for permission to reproduce extracts from SATs papers.

The publishers have made every effort to contact copyright holders but apologise if any have been overlooked.

Contents

Contents

Introduction

This workbook has been designed to be used alongside New National Framework Mathematics Core Book 9. However, it can also be used in conjunction with whatever resources are being used or as a fully stand alone resource. It is the ideal course companion to be used at home or at school. All topics from the Framework are covered in an accessible and stimulating format.

The book is divided into five sections. The first four relate to the strands of the National Numeracy Strategy Framework and contain 99 'topic pages'. Each page has been very carefully developed. At the top of each page are the core objectives to be covered. These are followed by a worked example and a summary of the key points. A range of differentiated questions follow, ranging from practice questions to games and puzzles.

The fifth section contains two SAT style practice papers, one for use with a calculator and one without.

Workbooks are also available for the Star and Plus books for all years.

1 Powers of ten

Let's look at ...
- understanding powers of ten
- recognising numbers in standard form
- prefixes associated with powers of ten

These are the points you need to know.

✓ This chart shows the **powers of ten**.

This is the base 10 number system.

Millions	Thousands			Hundreds						
Millions	Hundreds of thousands	Tens of thousands	Thousands	Hundreds	Tens	Units •	tenths	hundredths	thousandths
10^6	10^5	10^4	10^3	10^2	10^1	10^0 •	10^{-1}	10^{-2}	10^{-3}

We say $3 \cdot 5 \times 10^{-2}$ as 'three point five times ten to the negative two'.
$3 \cdot 5 \times 10^{-2}$ is written in **standard form**.

one digit before the decimal point

times ten to a power

✓ **Metric measures** are associated with the powers of ten.

Power	10^9	10^6	10^3	10^{-2}	10^{-3}	10^{-6}	10^{-9}	10^{-12}
Prefix	giga	mega	kilo	centi	milli	micro	nano	pico

Example 7 kilograms = 7×10^3 grams
1·5 nanometres = $1 \cdot 5 \times 10^{-9}$ metres
4 centilitres = 4×10^{-2} litres
20 microseconds = 20×10^{-6} seconds

(A) Write 10^{-4} in three **different ways**. _____ , _____ , _____

(B) *Is it standard?*

Which number in each list is **not** written in standard form?

a **A** $3 \cdot 4 \times 10^2$ **B** $1 \cdot 65 \times 10^5$ **C** $5 \cdot 8^4$ **D** $1 \cdot 9 \times 10^{-2}$ ☐

b **A** $0 \cdot 96 \times 10^3$ **B** $4 \cdot 04 \times 10^{-2}$ **C** $1 \cdot 91 \times 10^3$ **D** $9 \cdot 2 \times 10^{-7}$ ☐

(C) *Metric matters*

a Fill in the gaps. The first one is done for you.

  i 4 kilometres = $4 \times$ __10^3__ metres ii 562 millilitres = $562 \times$ ____ litres
  iii 11 nanoseconds = $11 \times$ ____ seconds iv 48 picograms = $48 \times$ ____ grams

b Complete these. The first one is done for you.

  i 2 megabytes = $2 \times$ __10^6__ bytes = $2 \times$ __1 000 000__ bytes = __2 000 000__ bytes
  ii 5 gigabytes = $5 \times$ _____ bytes = $5 \times$ _____ bytes = _____ bytes
  *iii 8 milliseconds = $8 \times$ _____ seconds = $8 \times$ _____ seconds = _____ seconds

c Match each measurement on the left with an equivalent on the right.

  i 7 millimetres •   • 7×10^3 metres   ii 5 milligrams •   • 5×10^3 grams
     7 centimetres •   • 7×10^{-3} metres   5 micrograms •   • 5×10^{-7} grams
     7 kilometres •   • 7×10^6 metres   5 kilograms •   • 5×10^4 grams
     7 megametres •   • 7×10^{-2} metres   50 kilograms •   • 5×10^{-3} grams
     7 nanometres •   • 7×10^{-9} metres   500 nanograms •   • 5×10^{-6} grams

How did you find this? EASY OK HARD

2 Multiplying and dividing by powers of ten

Let's look at ...
- multiplying and dividing by 10, 100, 1000, 10 000, ...
- multiplying and dividing by 0·1 and 0·01

These are the points you need to know.

✓ **Multiplying and dividing by powers of ten**
To **multiply by 10, 100, 1000, 10 000, ...** we move the digits one place to the left for each zero in 10, 100, 1000, 10 000, ...

✓ To **divide by 10, 100, 1000, 10 000, ...** we move the digits one place to the right for each zero in 10, 100, 1000, 10 000, ...

✓ **Multiplying by 0·1 is the same as dividing by 10.**
Multiplying by 0·01 is the same as dividing by 100. *Example* $43·45 \times 0·01 = 43·45 \div 100 = 0·4345$
Dividing by 0·1 is the same as multiplying by 10. *Example* $6·4 \div 0·1 = 6·4 \times 10 = 64$
Dividing by 0·01 is the same as multiplying by 100.

(A) The first man

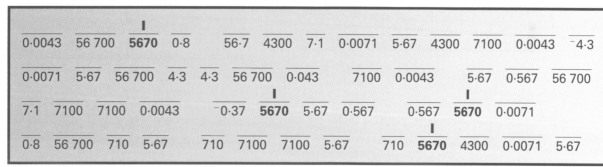

		5670										
0·0043	56 700	**5670**	0·8		56·7	4300	7·1	0·0071	5·67	4300	7100	0·0043	⁻4·3
0·0071	5·67	56 700	4·3	4·3	56 700	0·043		7100	0·0043		5·67	0·567	56 700
7·1	7100	7100	0·0043		⁻0·37	**5670**	5·67	0·567		0·567	**5670**	0·0071
0·8	56 700	710	5·67		710	7100	7100	5·67		710	**5670**	4300	0·0071	5·67

Write the letter beside each question above its answer in the box.

I $567 \times 10 = \mathbf{5670}$ **R** 43×100 **O** $7·1 \times 1000$ **T** $56·7 \div 10$ **D** $4·3 \div 100$

M $0·071 \times 100$ **A** $0·0567 \times 1000$ **N** $4·3 \div 1000$ **L** $8 \times 0·1$ **H** $5·67 \times 0·1$

P $0·43 \div 0·1$ **S** $0·71 \times 0·01$ **E** $567 \div 0·01$ **F** $7·1 \div 0·01$ **G** $^-43 \div 10$

W $^-3·7 \times 0·1$

(B) On holiday

a Peter bought 100 postcards at £0·28 each. How much did they cost altogether? _____

b One hundred passengers on a flight each had their luggage weighed. The total mass of the luggage was 1567 kg. On average, how heavy was each passenger's luggage? _____

c Richard bought a handmade Turkish rug. Its area was 3·75 m². What is this in cm²? _____

d Hanna's new suitcase had a capacity of 80 000 cm³. What is this in m³? _____

Use these facts to help with **c** and **d**.

$1 \, m^2 = 10\,000 \, cm^2$
$1 \, m^3 = 1\,000\,000 \, cm^3$

(C) Show time

Alyssa wrote:

$8·2 \times 0·1 = 8·2 \times \frac{1}{10}$
$= 8·2 \div 10$
So $8·2 \times 0·1$ is the same as $8·2 \div 10$

Use Alyssa's way to show that $9·75 \times 0·01$ is the same as $9·75 \div 100$.

How did you find this?   **EASY**   **OK**   **HARD**

3 Rounding

Let's look at ...
- **rounding to powers of ten and to decimal places**
- **using rounding to make estimates**

These are the points you need to know.

✓ We often **round** numbers when **estimating**.

Example A reporter might round 28 362, the number of people at a concert, to 28 000 (nearest thousand) or 30 000 (nearest ten thousand).

Example 342 × 78 is about 300 × 80 = 24 000 **or** 350 × 80 = 28 000.

✓ When **rounding to a given number of decimal places** we keep the digits we want. If the digit in the next decimal place is 5 or greater, we increase the last digit we are keeping by 1.

Examples 4·5362 to 1 d.p. = 4·5   17·038 to 1 d.p. = 17·0
26·3774 to 2 d.p. = 26·38   25·5 to 0 d.p. or the nearest whole number = 26

(A) Two tables

Complete these tables.

a

Number	Nearest 100	Nearest 1000	Nearest 10 000
8 321 423			
3 968 500			

b

Number	Nearest whole number	to 1 d.p.	to 2 d.p.
12·473			
19·987			

(B) Calculate it

a Find an approximate answer to these by writing down an estimate first.

   i 913 × 48 is about _____ × _____ = _____   **ii** 389 × 72 is about _____ × _____ = _____

b Use your calculator to solve this. Round your answer sensibly. Say what you have rounded to.

   Aaron's lounge is being re-carpeted.
   What is its area? _____

   4·72 m

   2·49 m

(C) Fun run

Sam heard that the number of participants in a fun run was 36 000 to the nearest thousand.

a What are the smallest and largest number of runners there could have been?
   smallest _____   largest _____

Later Sam read in the newspaper that there were 35 700 runners to the nearest 100.

b Write down three possible numbers of runners. _____, _____, _____

(D) South America

This gives the land area of the five largest countries in South America.

a Add the five land areas, then round the answer to the nearest 10 000 km². ____

b Round each land area to the nearest 10 000 then add them. ____

c Explain why the totals in **a** and **b** are different. ____

d The total land area of South America is about 17 840 000 km². Estimate the percentage of land which is **not** in the five largest countries. ____

*e The land area of the tenth largest country, Guyana, is 214 970 km². Estimate the percentage of land in South America which is taken up by the ten largest countries. ____

Land Area (km²)	
Brazil	8 511 965
Argentina	2 766 890
Peru	1 285 216
Colombia	1 138 910
Bolivia	1 098 581

You could take the mean of Bolivia and Guyana to estimate the size of the 6th, 7th, 8th and 9th largest countries.

How did you find this?   EASY   OK   HARD

4 Adding, subtracting, multiplying and dividing integers

Let's look at ...
- adding and subtracting positive and negative numbers
- multiplying and dividing positive and negative numbers

These are the points you need to know.

✓ **Adding a positive** number or **subtracting a negative** number is the same as **adding**.

Examples $\quad^-6 + {}^+4 = {}^-6 + 4 \qquad\qquad {}^-7 - {}^-4 = {}^-7 + 4$
$\qquad\qquad\quad = {}^-2 \qquad\qquad\qquad\qquad\quad = {}^-3$

✓ **Adding a negative** number or **subtracting a positive** number is the same as **subtracting**.

Examples $\quad 5 + {}^-3 = 5 - 3 \qquad\qquad {}^-4 - {}^+3 = {}^-4 - 3$
$\qquad\qquad\quad = 2 \qquad\qquad\qquad\qquad\quad = {}^-7$

✓ When we **multiply or divide with a positive and a negative number** we get a negative answer.

Examples $\quad^-6 \times 4 = {}^-24 \qquad\qquad 15 \div {}^-3 = {}^-5$

✓ When we **multiply or divide with two positive or two negative numbers** we get a positive answer.

Examples $\quad^-6 \times {}^-3 = 18 \qquad\qquad {}^-24 \div {}^-8 = 3 \qquad\qquad \frac{{}^-36}{{}^-4} = 9$

(A) Number chains

Complete each number chain.

a $\boxed{{}^-3} \xrightarrow{+2} \boxed{{}^-1} \xrightarrow{+3} \boxed{} \xrightarrow{-7} \boxed{} \xrightarrow{+3} \boxed{{}^-2} \xrightarrow{+10} \boxed{} \xrightarrow{+{}^-12} \boxed{} \xrightarrow{-4} \boxed{}$

b $\boxed{3} \xrightarrow{\times{}^-4} \boxed{} \xrightarrow{\div2} \boxed{} \xrightarrow{\div{}^-3} \boxed{} \xrightarrow{\times{}^-10} \boxed{} \xrightarrow{\div{}^-4} \boxed{} \xrightarrow{\times6} \boxed{} \xrightarrow{\div{}^-2} \boxed{}$

(B) What comes first?

Use the order of operations rules to find the answers to these.

a $4 + 2 \times 3 = $ _____

b $^-5 - \frac{8}{4} = $ _____

c $^-6(3 - 5) = $ _____

d $\frac{{}^-7 + {}^-11}{3} = $ _____

(C) Card games

Rebecca has these cards.

Use Rebecca's cards to answer these.

$\boxed{{}^-4} \ \boxed{7} \ \boxed{5} \ \boxed{10} \ \boxed{{}^-1} \ \boxed{2}$

a Make this true in two different ways. $\boxed{} - \boxed{} = 3 \qquad \boxed{} - \boxed{} = 3$

b Make this true in two different ways. $\boxed{} \times \boxed{} = {}^-20 \qquad \boxed{} \times \boxed{} = {}^-20$

c Use three cards so each of these have the **smallest** possible answer.

i $\boxed{} + \boxed{} + \boxed{} = $ ___

ii $\boxed{} + \boxed{} - \boxed{} = $ ___

(D) What are my numbers?

a Two numbers add together to give 2.
They multiply together to give $^-24$.
What are the two numbers? ____, ____

b Two numbers add together to give $^-9$.
They multiply together to give 14.
What are the two numbers? ____, ____

5 Prime factor decomposition

Let's look at ...
● highest common factors and lowest common multiples
● cancelling fractions using the HCF
● adding and subtracting fractions using the LCM

These are the points you need to know.

Remember

✓ The **highest common factor (HCF)** of two numbers is the largest factor common to both.

✓ The **lowest common multiple (LCM)** of two numbers is the smallest number that is a multiple of both.

Example $72 = 2^3 \times 3^2$
$60 = 2^2 \times 3 \times 5$

72 and 60 are written as products of their prime factors in index notation.

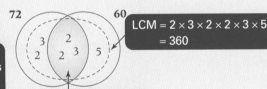

$LCM = 2 \times 3 \times 2 \times 2 \times 3 \times 5 = 360$

$HCF = 2 \times 2 \times 3 = 12$

✓ When **cancelling fractions**, we divide the numerator and denominator by their HCF.

Example $\frac{4\cancel{24}}{7\cancel{42}} = \frac{4}{7}$ Divide numerator and denominator by the HCF, 6.

✓ When **adding and subtracting fractions**, we first find the LCM of the denominators.

Example $\frac{5}{6} + \frac{4}{9} = \frac{15+8}{18}$ The LCM of 6 and 9 is 18. $\frac{5}{6} = \frac{15}{18}$ and $\frac{4}{9} = \frac{8}{18}$
$= \frac{23}{18} = 1\frac{5}{18}$

A Getting started

Complete the factor trees and table. Write each number as a product of prime factors in index notation.

a 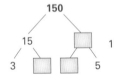 $150 =$ _____

b $550 =$ _____

c

2	504
2	252
2	

$504 =$ _____

B Cancel it

a Find the HCF of 8 and 14. _____ Use this to cancel $\frac{8}{14}$. _____

b Find the HCF of 15 and 21. _____ Use this to cancel $\frac{15}{21}$. _____

c Cancel these. **i** $\frac{45}{55}$ _____ **ii** $\frac{12}{42}$ _____ **iii** $\frac{63}{81}$ _____ **iv** $\frac{60}{400}$ _____

C Work it out

Calculate these by finding the LCM of the denominators.

a $\frac{1}{4} + \frac{2}{5} = \frac{\square+\square}{\square} = \frac{\square}{\square}$ **b** $\frac{2}{3} + \frac{1}{6} = \frac{\square+\square}{\square} = \frac{\square}{\square}$ **c** $\frac{5}{6} - \frac{3}{4} = \frac{\square-\square}{\square} = \frac{\square}{\square}$

*D Puzzles

a The HCF of 16 and another number is 4.
What might the other number be? Give three possible answers. ____, ____, ____

b Use each of the digits in the box once to make two two-digit numbers with an HCF of 8. ____, ____

2	6
5	7

How did you find this? **EASY** **OK** **HARD**

6 Common factors of algebraic expressions

Let's look at ...
- **finding highest common factors of algebraic expressions**

> These are the points you need to know.

✓ **Remember** • A factor is a number that divides exactly into another number.
• In algebra, letters stand for numbers.

✓ We can **find common factors of algebraic expressions**.

Example $2n = 2 \times n$
$3n = 3 \times n$
Both $2n$ and $3n$ can be divided exactly by n.
n is a common factor of $2n$ and $3n$.

Example $3bc^2$ and b^2c can both be divided exactly by b and by c.
bc is the highest common factor of $3bc^2$ and b^2c.

(A) Match it up

Match the two expressions on the left with their highest common factor on the right.

> Find the highest common factor that both can be divided by.

a
$3x$ and $7x$ • • x^2
$8x$ and $4x$ • • $4x$
$4x^2$ and x^2 • • x
$2x^2$ and $8x$ • • $3x$
$3xy$ and $6xz$ • • $2x$

b
$4ab$ and $5ac$ • • $2ab$
$8ac$ and $6ab$ • • b
$5a^2b$ and bc • • $3c$
$6a^2c$ and $3bc$ • • $2a$
$4ab^2$ and $6a^2b$ • • a

(B) Quick questions

Write down the highest common factor of each of these.

a $3y$ and $5y$ _____
b $7s$ and s _____
c $3x$ and x^2 _____

d $10a$ and 5 _____
e $3m$ and $12m$ _____
f $3qr$ and $6rs$ _____

g $4a^2b$ and ab^2 _____
h $7x^2yz$ and y^2z _____
i $15abc^2$ and $9bc$ _____

(C) Hunt the triple

This is a triple because the common factor of $9q^2$ and $6q$ is $3q$.

$3q$ is shaded because it is the common factor.

Find the triples in the grid.

Shade the common factor in each triple.

What is left over? _____

$5x$	$6x$	8	$3ab$	$6a^2$
x	$4n$	4	$3a$	$5a^2c$
$3x^2$	$21n$	$14n^2$	$7n$	$10c^2$
$9x$	$3x$	7	$6cn$	$5c$
$8xy$	xy	$21x^2y$	$2c$	$8c^2$

*(D) Missing expressions

a $3x^2$ and another expression have a common factor of $3x$.
What could the other expression be? Write down three possibilities. _____, _____, _____

b $4qr^2$ and another expression have a common factor of $2qr$.
What could the other expression be? Write down three possibilities. _____, _____, _____

c Two expressions have a common factor of $5a^2$. What could the expressions be?
Write down two possibilities. _____ and _____, _____ and _____

How did you find this? **EASY** **OK** **HARD**

7 Divisibility and algebra

Let's look at ...
● using algebra to prove some divisibility rules

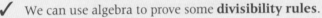

✓ We can use algebra to prove some **divisibility rules**.

Example   Any two-digit number in which the tens and units digits are the same is divisible by 11.
If the number of units is d, then the number of tens is $10d$.
The number is then $10d + d = 11d$.
$11d$ is always divisible by 11.

These are the points you need to know.

Ⓐ The nines

27, 63 and 81 are all two-digit numbers which are divisible by 9.

Carmel noticed that in all of these numbers the sum of the tens digit and the units digit is 9.

a   Find three more examples where this is true. _____, _____, _____

b   Carmel started to **prove** that two-digit numbers of this type are always divisible by 9.
Finish Carmel's proof.

> Let the units digit be u and the tens digit be t.
> The sum of the tens digit and the units digit is 9 so $t + u = 9$
>   or $u = 9 - t$.
>
> So the number is $10t + u = 10t + (9 - t)$
>
>   This means 10 times the tens digit.
>
> $= 10t + $ _____   remove brackets
> $= $ _____   collect like terms
> $= $ _____   factorise
> _____ is always divisible by 9.

*Ⓑ Tens and fives

a   Prove this statement.
Any two-digit number is divisible by 10 if the units digit is 0.

> Let the tens digit be t and the units digit is 0.
> The number is $10t + 0$.
> _____

b   Prove this statement.
Any two-digit number is divisible by 5 if the units digit is 5.

> Let the tens digit be t and the units digit is 5.
> The number is _____ = _____ .
> _____

c   Prove this statement.
Any three-digit number is divisible by 5 if the units digit is 5.

> Let the hundreds digit be h, the tens digit be t and the units digit is 5.
> The number is $100h + 10t + 5 = 5($_____$)$
> _____

How did you find this?   [EASY] [OK] [HARD]

8 Powers

Let's look at ...
● understanding and using index notation

These are the points you need to know.

✓ **Remember** ● We write $3 \times 3 \times 3 \times 3$ as 3^4.
● The 4 in 3^4 is called an **index**.

The plural of index is **indices**.

● On a calculator, squares are keyed using $\boxed{x^2}$, cubes are keyed using $\boxed{x^3}$.

✓ On a calculator indices are keyed using $\boxed{\wedge}$.

Example 32^5 is keyed as $\boxed{32}\boxed{\wedge}\boxed{5}\boxed{=}$ to get 33 554 432.

✓ Any number to the power of zero equals 1. $\quad x^0 = 1$

(A) Calculator time

Use your calculator to find the value of each of these.

Give the answers to d and f to 2 d.p.

a 16^3 _____ **b** $2 \cdot 7^2$ _____ **c** 4^6 _____ **d** $7 \cdot 1^4$ _____

e $(^-7)^3$ _____ **f** $(^-1 \cdot 9)^6$ _____ ***g** $(\frac{1}{5})^3$ _____ ***h** $8^2 \times 5^4$ _____

(B) Mixed bag

a Find the value of each of these.

i 8^0 _____ **ii** $(^-1 \cdot 9)^0$ _____ **iii** y^0 _____

b Complete the following.

i $100 = 10^\square$ **ii** $10 = 10^\square$ **iii** $0 \cdot 001 = 10^\square$ **iv** $\frac{1}{10} = 10^\square$ **v** $1 = 10^\square$

c 64 can be written in index form in four different ways. Two ways are 8^2 and 64^1.
Find the other two ways. _____, _____

d What power of 5 has a value of 625? _____

e What power of 5 multiplied by a power of 2 equals 400? _____ × _____ = 400

f $2^3 + 3^3 = 35$
Write 65 as the sum of two cubes. _____ + _____ = 65

(C) True or false?

Decide whether each of the following is true (T) or false (F).

a $2^2 + 2^2 = 2^6$ _____ **b** $(3 + 4)^2 = 3^2 + 4^2$ _____ **c** $(9 - 7)^2 = 9^2 - 7^2$ _____

d $(15 + 10)^0 = 15^0 + 10^0$ _____ **e** $4^2 = 2^4$ _____ **f** $4^3 = 3^4$ _____

g $25^2 = 5^4$ _____

*(D) Card games

Andrew has these number cards.

a Which card has the largest number on it? $\boxed{}$

b Which card is equal to 2^8? $\boxed{}$

Penny has these number cards.

c Circle Penny's cards which have square numbers on them.

---

9 Multiplying and dividing numbers with indices

Let's look at ...
● using the index laws

These are the points you need to know.

✓ **The index laws**
When multiplying, add the indices.

Example $5^4 \times 5^7 = 5^{(4+7)} = 5^{11}$

When dividing, subtract the indices.

Example $\frac{9^{12}}{9^5} = 9^{(12-5)} = 9^7$

The base numbers must be the same.

(A) What's missing?

Fill in the gaps.

a $3^2 \times 3^4 = (\underline{} \times \underline{}) \times (\underline{} \times \underline{} \times \underline{} \times \underline{})$
$= 3\underline{}$

b $7^3 \times 7^5 = (\underline{}) \times (\underline{})$
$= \underline{}\underline{}$

c $4^5 \div 4^3 = \dfrac{\underline{} \times \underline{} \times \underline{} \times \underline{} \times \underline{}}{\underline{} \times \underline{} \times \underline{}}$
$= 4\underline{}$

d $6^6 \div 6^5 = \dfrac{\underline{}}{\underline{}}$
$= \underline{}$

(B) That's high

Simplify each of these and shade the answer in the grid.
The shaded squares will make a number which completes the sentence below.

4^{56}	4^5	5^3	6^9	5^5	5^{11}
5^1	6^{12}	6^5	6^6	4^6	5^{24}
5^2	4^3	4^2	5^9	4^8	5^{10}
5^{25}	5^0	6^4	4^{10}	6^3	6^8
6^1	6^2	6^7	4^7	6^{16}	4^{15}

a $4^3 \times 4^2 = \mathbf{4^5}$
b $4^1 \times 4^6$
c $4^7 \times 4^8$
d $4^9 \div 4^6$
e $6^3 \times 6^3$
f $6^{12} \div 6^4$
g $\frac{6^8}{6^6}$
h $\frac{6^{20}}{6^4}$
i $5^5 \times 5^5$
j $\frac{5^5}{5^5}$
k $5^2 \times 5^3 \times 5^4$
l $5^0 \times 5^7 \times 5^4$
*m $\frac{4^5 \times 4^6}{4^3}$
*n $\frac{5^{12}}{5^2 \times 5^5}$
*o $\frac{6^{12} \times 6^4}{6^3 \times 6^1}$
*p $\frac{6^8 \times 6^7}{6^2 \times 6^4}$

Due to Earth's gravity it is impossible for mountains to be higher than ____ km.

(C) Odd one out

Circle the odd one out in each row.

a $2^2 \times 2^3$ 2^6 32 2^5

b 4^4 $4^{12} \div 4^3$ $\frac{4^{12}}{4^3}$ 4^9

c $7^2 \times 7^0 \times 7^3$ $7^0 \times 7^3$ 7^5 $7^2 \times 7^3$

d $\frac{5^7}{5^2 \times 5^3}$ $5^8 \div 5^6$ $5^2 \times 5^0$ $\frac{5^9}{5^3 \times 5^3}$

(D) Cameron's idea

Cameron thinks that $5^3 \times 2^4 = 10^7$. Is he correct? _____ Explain why or why not. _____

_____

*(E) Make it simple

Use the index laws to simplify these.

a $a^4 \times a^3$ ____
b $y^8 \times y^2$ ____
c $p^6 \div p^4$ ____
d $r^8 \div r^3$ ____
e $\frac{x^7}{x^4}$ ____
f $\frac{z^9}{z^2}$ ____
g $b^2 \times b^3 \times b^4$ ____
h $d^5 \times d^0 \times d^6$ ____

How did you find this?   **EASY**   **OK**   **HARD**

10 Square roots and cube roots

 except for B and D

Let's look at ...
- **finding square roots and cube roots**
- **properties of square roots and cube roots**

These are the points you need to know.

✓ **Remember**
- $\sqrt{81}$ means the **positive square root** of 81, which is 9.
- $\pm\sqrt{81}$ means the **positive** and **negative** square roots of 81, which are 9 and $^-9$. $9 \times 9 = 81$ and $^-9 \times {}^-9 = 81$
- $\sqrt[3]{27}$ means the **cube root** of 27. $\sqrt[3]{27} = 3$ because $3 \times 3 \times 3 = 27$.

✓ On a calculator we use $\boxed{\sqrt[3]{}}$ to find **cube roots**.
The cube root of a positive number is positive.
The cube root of a negative number is negative.
  Example $\sqrt[3]{216}$ is keyed as $\boxed{\sqrt[3]{}}\,\boxed{216}\,\boxed{=}$ to get 6. $\sqrt[3]{^-27}$ is keyed as $\boxed{\sqrt[3]{}}\,\boxed{(-)}\,\boxed{27}\,\boxed{=}$ to get $^-3$.

✓ We can estimate the value of square roots and cube roots using **trial and improvement**.
  Example 19 lies between the two square numbers, 16 and 25, so $\sqrt{19}$ lies between $\sqrt{16}$ and $\sqrt{25}$.
    $\sqrt{19}$ lies between 4 and 5.
      Try 4·3 $4·3 \times 4·3 = 18·49$ too small
      Try 4·4 $4·4 \times 4·4 = 19·36$ too big
      We now know that $\sqrt{19}$ lies between 4·3 and 4·4.
      We can continue by trying 4·32 or 4·35, for example.

(A) Use your head

a Find these.
  i $\sqrt{9}$ ____   **ii** $\pm\sqrt{25}$ _____   **iii** $\pm\sqrt{64}$ _____   **iv** $\sqrt{121}$ ____   **v** $\sqrt[3]{8}$ ____

b Complete these sentences with **positive** or **negative**.
  i The cube root of a _____ number is positive.
  ii The cube root of a _____ number is negative.

c Vanessa found the square root of 196 by factorising.
  $\sqrt{196} = \sqrt{4 \times 49} = \sqrt{4} \times \sqrt{49} = 2 \times 7 = 14$

Hint: Try dividing by square numbers like 4 and 9.

  Find these square roots by factorising. Show your working.
  i $\sqrt{225}$
  ii $\sqrt{324}$

(B) Calculator time

Use your calculator to find these. Give the answers to 2 d.p.
  a $\sqrt{478}$ ____   **b** $\sqrt[3]{13}$ ____   **c** $\sqrt{98 - 39}$ ____   **d** $\sqrt[3]{19^2 - 9^2}$ ____   **e** $\sqrt{17} + \sqrt{41}$ ____

(C) Find the truth

Circle the **true** statement or statements.

$\sqrt{4} + \sqrt{36} = \sqrt{40}$ $\sqrt{4} \times \sqrt{25} = \sqrt{100}$ $\sqrt{16} - \sqrt{4} = \sqrt{12}$ $\sqrt{9} \times \sqrt{9} = \sqrt{81}$

(D) Trial and improvement

Estimate the value of $\sqrt{40}$ to 2 d.p. using trial and improvement.
Do not use the $\boxed{\sqrt{}}$ key. Show your working.

See the example in the box at the top.

11 Multiplying and dividing by numbers between 0 and 1

Let's look at ...
● the result of multiplying or dividing a number by another number which is between 0 and 1

These are the points you need to know.

✓ When we multiply **a positive number**, a, **by a number between 0 and 1**, the answer is **smaller** than a.

✓ When we **divide a positive number**, b, **by a number between 0 and 1**, the answer is **larger** than b.

Examples $1 \cdot 56 \times 0 \cdot 7 = 1 \cdot 092$ The answer is less than 1·56 because we are multiplying by a number between 0 and 1
   $1 \cdot 56 \div 0 \cdot 4 = 3 \cdot 9$ The answer is greater than 1·56 because we are dividing by a number between 0 and 1

(A) *Choose carefully*

Circle the calculations which will have an answer of less than 50.
Do not calculate the answers.

$50 \times 0 \cdot 3$ $50 \div 0 \cdot 7$ $50 \times 0 \cdot 9$ $50 \div 28$ $50 \div \frac{1}{2}$ $50 \div 1 \cdot 6$

$50 \times 0 \cdot 04$ $50 \times 1 \cdot 5$ $50 \div 2\frac{1}{4}$ $50 \times 5\frac{1}{3}$ $50 \div 0 \cdot 001$

(B) *Maze time*

The correct path through this maze goes through calculations which have an answer of less than 0·08. Shade the correct path. The first step has been done for you.

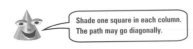

Shade one square in each column. The path may go diagonally.

START						FINISH
$0 \cdot 08 \div \frac{1}{2}$	$0 \cdot 08 \div 0 \cdot 1$	$0 \cdot 08 \div 0 \cdot 5$	$0 \cdot 08 \times 0 \cdot 09$	$0 \cdot 08 \times 1 \cdot 9$	$0 \cdot 08 \times \frac{9}{8}$	
$0 \cdot 08 \times \frac{1}{2}$	$0 \cdot 08 \times 10$	$0 \cdot 08 \div 5$	$0 \cdot 08 \div 0 \cdot 09$	$0 \cdot 08 \div 1 \cdot 9$	$0 \cdot 08 \times \frac{8}{9}$	
$0 \cdot 08 \times 2$	$0 \cdot 08 \times 0 \cdot 1$	$0 \cdot 08 \div 0 \cdot 05$	$0 \cdot 08 \times 90 \cdot 9$	$0 \cdot 08 \div 0 \cdot 9$	$0 \cdot 08 \times 8 \cdot 9$	

(C) *Patterning*

a   Complete these patterns.

i $6 \times 30\,000 = 180\,000$
   $6 \times 3000 = 18\,000$
   $6 \times 300 = $ _____
   $6 \times 30 = $ _____
   $6 \times 3 = $ _____
   $6 \times 0 \cdot 3 = $ _____
   $6 \times 0 \cdot 03 = $ _____

ii $24\,000 \div 6000 = 4$
   $24\,000 \div 600 = 40$
   $24\,000 \div 60 = 400$
   $24\,000 \div 6 = $ _____
   $24\,000 \div 0 \cdot 6 = $ _____
   $24\,000 \div 0 \cdot 06 = $ _____

b   Use the patterns to find the answers to these.

i $6 \times 0 \cdot 003 = $ _____   **ii** $24\,000 \div 0 \cdot 0006 = $ _____

(D) *What's missing?*

a   What goes in the gap, **greater** or **less**?

i   When we **divide** a positive number, c, by a number between 0 and 1, the answer will be _____ than c.

ii   When we **multiply** a positive number, d, by a number between 0 and 1, the answer will be _____ than d.

b   What goes in the gap × or ÷? Do not do the calculation.

i $2 \cdot 8 \ \square \ 0 \cdot 4 = 7$ **ii** $0 \cdot 9 \ \square \ 0 \cdot 84 = 0 \cdot 756$ **iii** $12 \ \square \ \frac{5}{6} = 10$

How did you find this?   **EASY**   **OK**   **HARD**

12 Order of operations

Let's look at ...
● carrying out operations in order

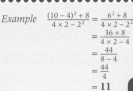

These are the points you need to know.

✓ The **order of operations is**   **B**rackets   **Remember** **BIDMAS**.   *Example* $\frac{(10-4)^2+8}{4\times2-2^2}=\frac{6^2+8}{4\times2-2^2}$
   Indices
   Division and **M**ultiplication $=\frac{36+8}{4\times2-4}$
   Addition and **S**ubtraction $=\frac{44}{8-4}$
                                               $=\frac{44}{4}$
                                               $=11$

Work out the whole numerator and the whole denominator first.

✓ We can work out simple calculations mentally and more difficult ones using a calculator.

A Mathematicians from history

Find the answers to these mentally.
Put the answers in order from smallest to largest, with their letters beside them.
The letters will each spell a famous mathematician's name.

1 **E** $4+6\times7$ _____   **E** $30\div(7-2)$ _____   **U** $\frac{63-7}{8}$ _____
   L $(8-3)^2+13$ _____   **R** $(^-6)^2+5\times3$ _____   The mathematician is _____.

2 **A** $20+4^2\times2$ _____   **L** $3(7-2)^2-20$ _____   **S** $\frac{8^2-10}{2}$ _____   **P** $\frac{36}{(7-4)^2}$ _____
   A $\frac{(4+6)^2}{5\times2}$ _____   **C** $(5+2)^2-(^-3)^2$ _____
   The mathematician is _____.

3 **E** $\sqrt{81-32}$ _____   **V** $\sqrt[3]{112-48}$ _____   **I** $\sqrt{4}+3\times7$ _____
   T $\sqrt{9^2+63}$ _____   **A** $(5\times4-4\times3)^2$ _____   The mathematician is _____.

B What's the difference?

Explain the difference between A and B.

a **A** $(3+5)^2$ **B** $3+5^2$ _____

_____

b **A** $\sqrt{16+9}$ **B** $\sqrt{16}+9$ _____

_____

C What's missing?

Put +, −, × or ÷ into these calculations to make the answer correct.

a $7\ \square\ (2\ \square\ 3)=35$ **b** $(4\ \square\ 3)^2\ \square\ 10\ \square\ 3=19$ **c** $(9\ \square\ 6)\ \square\ (7\ \square\ 2)=3$

D Puzzles

a My number is between 70 and 80.
   My number can be made using the numbers 2, 5 and 7, the operations +, −, ^{2} just once each and one set of brackets.
   What is my number? _____.

*b Use the digits 1, 3, 5 and 7, the operations −, ×, +, ^{2} and two sets of brackets to get the answer 15.
   Use each digit and each operation just once. _____.

13 Mental calculations

Let's look at ...
- adding, subtracting, multiplying and dividing mentally
- calculating with fractions, decimals and percentages mentally

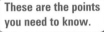
These are the points you need to know.

✓ We can **add and subtract** mentally using complements, partitioning, counting up, nearly numbers, compensation or facts we already know.

✓ We can **multiply and divide** mentally using place value, partitioning, factors, near tens, known facts or doubling and halving.

✓ We can **convert between fractions, decimals and percentages mentally**.
We can also do simple fraction, decimal and percentage calculations mentally.
You need to know

$$\frac{1}{8} = 0·125 = 12\frac{1}{2}\% \qquad \frac{1}{3} = 0·\dot{3} = 33\frac{1}{3}\% \qquad \frac{2}{3} = 0·\dot{6} = 66\frac{2}{3}\%$$

Examples
$$\begin{aligned}\frac{3}{8} &= 3 \times \frac{1}{8}\\ &= 3 \times 0·125\\ &= 0·375 \text{ as a decimal}\\ &= 37\frac{1}{2}\% \text{ as a } \%\end{aligned}$$

$$\begin{aligned}\frac{2}{3} \text{ of } 45·3 &= \frac{2}{3} \times 45·3\\ \frac{1}{3} \times 45·3 &= 45·3 \div 3\\ &= 15·1\\ \frac{2}{3} \times 45·3 &= 2 \times 15·1\\ &= \mathbf{30·2}\end{aligned}$$

To find 17·5% of 120:
10% of 120 = 12
5% of 120 = 6
2·5% of 120 = 3
so 17·5% of 120 = 12 + 6 + 3
= 21

(A) Quick questions

a Subtract five from the square of six. _____
b What number is nine less than four? _____
c 13, 7, 1, ⁻5, ... What comes next? _____
d 5x = 70. What is x? _____
e Express $\frac{3}{4}$ as a decimal. _____
f Convert 0·6 to a fraction. _____
g Express $\frac{2}{5}$ as a percentage. _____
h Convert 1·145 to a percentage. _____
i Increase £50 by 10%. _____
j Decrease 300 mm by 15%. _____

(B) Market day

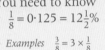

| $\overline{920}$ | $\overline{7·3}$ | | $\overline{112}$ | $\overline{8·72}$ | $\overline{7·5}$ | | $\overline{50}$ | $\overline{920}$ | $\overline{36}$ | $\overline{36}$ | $\overline{2·8}$ | $\overline{7·5}$ | | $\overline{8}$ | $\overset{G}{\overline{100}}$ | $\overline{7·5}$ | $\overline{0·2}$ |

| $\overline{0·2}$ | $\overline{45}$ | $\overset{G}{\overline{100}}$ | $\overline{8}$ | $\overline{247}$ | | $\overline{42}$ | $\overline{90}$ | $\overline{0·2}$ | $\overline{112}$ | | $\overline{7·3}$ | $\overline{920}$ | $\overline{7·3}$ | $\overline{7·5}$ |

| $\overline{112}$ | $\overline{920}$ | $\overline{50}$ | $\overline{7·5}$ | $\overline{0·2}$ | | $\overline{8}$ | $\overline{0·2}$ | | $\overline{50}$ | $\overline{45}$ | $\overline{42}$ | $\overline{8·72}$ | | $\overline{8}$ | $\overline{0·2}$ | | $\overline{50}$ | $\overline{920}$ | $\overline{2·8}$ | $\overline{500}$ |

Write the letter beside each question above its answer in the box.

G 27 + 73 = **100**   N 8·2 − 0·9   A 4·56 + 3·44   H 12 − 3·28   M 4500 ÷ 90   I 230 × 4
L 4 × 0·7   S 0·8 ÷ 4   R 13 × 19   O 45 ÷ 0·5   D $\frac{3}{4}$ of 48   U $\frac{5}{7}$ of 63
T $1\frac{2}{5}$ of 80   E 30% of 25   K 125% of 400   C $17\frac{1}{2}\%$ of 240

(C) Sale time

Find the sale price of each of these items.

a
Was £240
25% off
Sale price = _____

b
Was £62
30% off
Sale price = _____

c
Was £25
35% off
Sale price = _____

(D) Puzzle

a Complete these.
  i 8 is halfway between 4·6 and _____.
  ii 8 is halfway between ⁻3 and _____.
b What number is halfway between 47 × 43 and 53 × 43? _____

How did you find this?   EASY   OK   HARD

14 Solving word problems mentally

Let's look at ...
● applying mental skills to solving problems

Here is an example to show you what to do.

✓ We can solve **word problems mentally**.

Example Mel bought a new coat in this sale.
Its original price was £95.
The new price can be worked out mentally.

A 20% reduction is the same as 80% of the original price.
We need to find 80% of £95.
  10% of £95 = £9·50
so 80% of £95 = 8 × £9·50
         = 2 × 2 × 2 × £9·50
         = 2 × 2 × £19
         = 2 × £38
         = **£76**

SALE

all coats reduced by **20%**

A Quick questions

Answer these questions as quickly as possible.

a How many edges has a cylinder? _____

b How many hours are there in $2\frac{1}{4}$ days? _____

c What is the volume of a cube, if each side is 5 cm long? _____

d Two angles in a triangle are 42° and 79°. What is the third angle? _____

e Cara's mass went from 42·3 kg to 48·1 kg in a year. How much mass did she gain? _____

f What is the area of a rectangle with sides 4·3 cm and 20 cm? _____

g A train travels at an average speed of 120 km/h. How far does it travel in 45 minutes? _____

h On a map 1 cm represents 5 km. A road is 70 km long, how long is it on the map? _____

i In a class of 30 students 12 are boys. What is this as a percentage? _____

j The ratio of cows to sheep on a farm is 3 : 7. There are 75 cows, how many sheep are there?
_____

B Take your time

Answer these questions carefully. You may like to use jottings.

BOOK SALE
5 BOOKS FOR £19
or £4·60 each

a Guy wanted to buy four books. How much more would it cost him to buy five books? _____

b The probability that it will snow in Gavin's home town on Christmas Day is 0·05. If Gavin is 60 years old, how many snowy Christmases would you expect him to have had? _____

c Stephen bought some magazines and some CDs at a market. The magazines cost £3 each and the CDs cost £7 each. Stephen spent £32.
How many magazines did he buy? _____

d A vet recorded the number of kittens 60 cats had in a year.
25% had 6 kittens, 30% had 5 kittens and the rest had 4 kittens.
In total, how many kittens were born? _____

*e Five women shared the winnings of a lottery equally between them.
Belinda gave away $\frac{3}{10}$ of her share. If Belinda gave away £63, what was the total lottery win?
_____

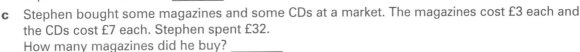

How did you find this? **EASY** **OK** **HARD**

15 Estimating

 except f

Let's look at ...
● deciding the degree of accuracy needed
● making estimates

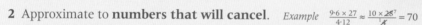

✓ When estimating, the **degree of accuracy** needed depends on the situation.

Examples ● A reporter only needs a rough estimate of the number of people at a football match.
● The organisers of a camp need to know the exact number coming.

> ≈ means approximately equal to.

These are the points you need to know.

✓ Use these guidelines when estimating answers.

1 Approximate to **'nice' numbers** that are easy to work with.

Example $61 \cdot 6 \div 7 \approx 63 \div 7 = 9$ rather than $60 \div 7$.

2 Approximate to **numbers that will cancel.** *Example* $\frac{9 \cdot 6 \times 27}{4 \cdot 12} \approx \frac{10 \times 28^7}{\cancel{4}} = 70$

3 When **multiplying**, try to **round one number up and one number down**.

Example $12 \cdot 5 \times 6 \cdot 5 \approx 12 \times 7 = 84$

4 When **dividing**, try to **round both numbers up or both numbers down**.

Example $\frac{47 \cdot 2}{5 \cdot 6} = \frac{48}{6} = 8$ 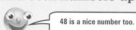 48 is a nice number too.

(A) How accurate?

Decide on the degree of accuracy needed for these.
Choose A, B or C from the box.

> **A** As accurate as possible.
> **B** A rough estimate will do.
> **C** An estimate is fine but it must be reasonably accurate.

a A motel wants to know what time Mr Scott will arrive. _____

b A doctor asks her patient how much exercise he does in a week. _____

c A dressmaker measures the waist and hips of a woman having her wedding dress made. _____

d A reporter asks the organisers how many people attended a variety show. _____

e An insurance company needs to know the cost of repairs to a car before paying the bill. _____

(B) Who's the best?

Decide whose estimate is the best for each of these.

Question	Joy	Gulam	Karen	Best estimate
a $9 \cdot 5 \times 10 \cdot 5$	9×10	10×11	10×10	
b $35 \cdot 4 \div 8 \cdot 7$	$36 \div 9$	$35 \div 9$	$36 \div 8$	
c $29 \cdot 1 \div 6 \cdot 6$	$29 \div 6$	$30 \div 7$	$30 \div 6$	

(C) Your turn

Estimate the answers to these. Show your working. The first one is started for you.

a $53 \times 49 \approx$ __50__ $\times$ __50__ $=$ _____

b $6 \cdot 38 \times 10 \cdot 61 \approx$ ____ $\times$ ____ $=$ ____

c $62 \cdot 8 \div 7 \cdot 9 \approx$ ____ $\div$ ____ $=$ ____

d $\frac{8 \cdot 13 + 3 \cdot 68}{4 \cdot 4} \approx \frac{\_\_\_ + \_\_\_}{\_\_\_} = \_\_\_$

(D) Wordy wonderings

Estimate, then use your calculator, to find the answers to the following. If rounding is needed, round your answers sensibly.

a A school hall is 11·6 m wide and 51·2 m long. Find the area of the floor.
Estimate = _____   Accurate answer = _____

b One pound is 16 ounces. Baby Joshua weighs 309·4 ounces. How many pounds is this?
Estimate = _____   Accurate answer = _____

How did you find this? EASY   OK   HARD

16 Written calculation

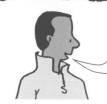

Let's look at ...
● using written methods for addition, subtraction, multiplication and division, with whole numbers and decimals

These are the points you need to know.

✓ We always **estimate** the answer first.

✓ We can **check answers** by checking the answer is sensible, checking the answer against our estimate, working the problem backwards, doing an equivalent calculation or checking the last digits.

To **divide by a decimal** we do an equivalent calculation and divide by a whole number.

Example   Calculate 248 ÷ 3·2.

$248 ÷ 3·2 ≈ 240 ÷ 3 = 80$ ◄

240 is a 'nice' number
$\frac{248 × 10}{3·2 × 10} = \frac{2480}{32}$

248 ÷ 3·2 is equivalent to 2480 ÷ 32

$32) 2480$
   $\underline{-2240}$ **32 × 70**
      240

80 is our estimate.

   $\underline{-224}$ **32 × 7**
    16
   $\underline{-16}$ **32 × 0·5**

Answer   **77·5**

 Odd one out

You will need some extra paper for your working for (A), (C), (D) and (E).

Find the answer to each question in the row, and circle the odd one out.

a   5827·3 – 72·4 – 3·06 _____   7301 – 1469·91 – 89·25 _____   5689·1 + 52·35 + 0·39 _____

b   460 × 2·8 _____   368 × 3·5 _____   225·8 × 5·7 _____

c   8·5 × 0·63 _____   81 ÷ 15 _____   91·8 ÷ 17 _____

B **Take your pick**

a   Which is equivalent to the calculation given?

  i   491 ÷ 0·3   **A** 491 ÷ 3   **B** 4910 ÷ 30   **C** 4910 ÷ 3   **D** 49 100 ÷ 3   _____
  ii 36·8 ÷ 0·4   **A** 3·68 ÷ 4   **B** 368 ÷ 4   **C** 3680 ÷ 4   **D** 368 ÷ 40   _____
  iii 0·056 ÷ 9·2   **A** 56 ÷ 92   **B** 56 ÷ 920   **C** 5·6 ÷ 92   **D** 0·56 ÷ 92   _____

b   Without using a calculator, pick a possible answer to the calculation. Explain your choice.
   317 × 2·2   **A** 583·4   **B** 697·4   **C** 678·2 _____ _____

C **Decimal division**

Give the answers to these to 1 d.p.

a   713 ÷ 0·3 _____   **b**   195 ÷ 0·7 _____   **c**   81·4 ÷ 0·06 _____   **d**   53·9 ÷ 2·9 _____

D **Gavin's gardens**

This is a plan of Gavin's backyard.
Calculate these.

a   The width of the flower bed. _____
b   The area of the patio. _____
c   The length of the vegetable garden. _____
d   What is the perimeter of Gavin's vegetable garden? _____

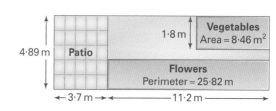

4·89 m | Patio
1·8 m   Vegetables   Area = 8·46 m²
Flowers   Perimeter = 25·82 m
← 3·7 m → ← 11·2 m →

---

17 Choosing a strategy for calculation

Let's look at ...
● solving problems using a mental, written or calculator method

These are the points you need to know.

✓ We can choose a **mental, written or calculator** method to do calculations.
For each of these questions use a mental method if possible.
Only use a calculator if you need to.

(A) *At the mall*

a Jennie bought a shirt. She gave the shop assistant £40 and got £11·75 in change. How much did the shirt cost? _____

b If 22 pencils cost £3·96, how much does one cost? _____

c A wrought-iron candlestick costs £15·30, and candles cost 95p each. How much would it cost Vince to buy a pair of candlesticks and 12 candles? _____

(B) *'The Golden Sword'*

Class 9N put on a musical. They charged £1·50 admission. After the musical Alastair sorted the admission money into piles and counted it. This table shows the money that was collected.

How many people paid to watch the play? _____

Money	£5 notes	£2 coins	£1 coins	50p coins	20p coins	10p coins
Frequency	3	28	53	42	17	16

(C) *Moving time*

Robina and John are moving house. They want to hire a truck for a day to move all of their furniture to their new house. They estimate that they have 60 m^3 of furniture. A return trip to their new house is 50 km long.

These tables show the charges for a smaller 20 m^3 capacity truck and a larger 30 m^3 capacity truck.

Which truck would be cheaper for Robina and John to use? _____
How much cheaper? _____

20 m^3 truck	
Cost per day	£60
Cost per km	40p
Insurance	£11

30 m^3 truck	
Cost per day	£80
Cost per km	45p
Insurance	£11

*(D) Rock-fall

A mine worker wants to clear a rock-fall. He has a stick of dynamite with a fuse which is 19·8 m long.
The fuse burns at a rate of 12 cm per second.
The dynamite will explode instantly when the burning fuse reaches it.

END

How long will it take for the dynamite to explode if

a the worker lights the very end of the fuse? ____ mins ____ secs
b the worker lights the fuse 6·24 m from the end? ____ mins ____ secs

*(E) Stationery sale

Three pencils and two pens cost £1·51.
Two pencils and three pens cost £1·59.
How much does a pencil cost? _____

How did you find this? **EASY** **OK** **HARD**

18 Using a calculator

Let's look at ...
● using a variety of keys on a calculator

These are the points you need to know.

✓ On a **calculator** we use

π

(Shift) (EXP) for π (−) to enter a negative number

(x^2) to square a number   (x^3) to cube a number

(^) to enter numbers with indices   (√) to find the square root

($a^{b/c}$) to enter fractions   (() and ()) for brackets

(STO) (M+) to store a number in memory   (0) (STO) (M+) to clear the memory.

Examples   To find $4{\cdot}62^4 \times 3{\cdot}1^5$

key (4·62) (^) (4) (×) (3·1) (^) (5) (=) to get 130 429·66 (2 d.p.).

To find $2\frac{1}{3} + 4\frac{5}{8}$

key (2) ($a^{b/c}$) (1) ($a^{b/c}$) (3) (+) (4) ($a^{b/c}$) (5) ($a^{b/c}$) (8) (=) to get $6\frac{23}{24}$.

(A) Whoops!

Shaun keyed in each of these calculations incorrectly. Explain what his mistakes are.

a $3{\cdot}17^3 \times 6{\cdot}2^4$ (3·17) (x^3) (×) (6·2) (x^2) (4) (=) _____

_____

b $4\frac{2}{5} + \frac{6}{7}$ (4) (2) ($a^{b/c}$) (5) (+) (6) ($a^{b/c}$) (7) (=) _____

_____

c $\frac{17}{12 + 4(5-2)}$ (17) (÷) (() (() (12) (+) (4) (×) (() (5) (−) (2) ()) (=) _____

_____

d $\sqrt{7^3 - \pi}$ (√) (() (7) (x^3) ()) (−) (Shift) (π) _____

_____

(B) Crossnumber

Complete this crossnumber. Write the answers to 2 d.p. when necessary.
The first one is done for you:

Across

1 $\frac{536}{0{\cdot}65 \times 4}$

5 $9{\cdot}8^2 - 9{\cdot}04$

6 $4{\cdot}75 \times \sqrt{214 + 186}$

7 $^-195{\cdot}46 \div 2{\cdot}9$

9 $19^2 - 8^3 + 471$

10 $\frac{(16+5)^2 + (19-3)^2 + 23}{4(18-3)}$

11 $\frac{(13-8)^2 - 1}{3(4{\cdot}9 - 3{\cdot}1)}$

Down

1 $\frac{143 + 51}{7{\cdot}2}$

2 $\frac{60 \times 80}{3+5}$

3 $\frac{12{\cdot}8 + 4 \times 7}{3}$

4 $\sqrt{17^2 + 4^2}$

* 7 $^-(97 \times 4 + 102) + 3 \times 48 - (567 + 136)$

8 $\frac{20{\cdot}7 - 1{\cdot}2 \times 0{\cdot}9}{0{\cdot}26}$

9 $3^5 + 9^2$

Take up a whole square for negative signs and decimal points.

(C) Keying fractions

Find the answer to these using your calculator.

a $4\frac{1}{3} + 2\frac{3}{5}$ _____   **b** $6\frac{2}{7} - 4\frac{3}{4}$ _____   **c** $11\frac{1}{9} \times 2\frac{3}{5}$ _____

d $1\frac{2}{3} \div \frac{7}{8}$ _____   **e** $^-3\frac{1}{2} - 5\frac{6}{7}$ _____   **f** $^-8\frac{8}{9} \times 4\frac{3}{7}$ _____

How did you find this?   (EASY)   (OK)   (HARD)

19 Fractions, decimals and percentages

Let's look at ...
- expressing proportions as fractions, decimals and percentages
- converting between fractions, decimals and percentages

These are the points you need to know.

✓ We can express **proportions** using **fractions, decimals and percentages**.

✓ We can **convert between fractions, decimals and percentages**.

Example To write $\frac{7}{20}$ as a percentage either
a write with denominator of 100 **or** **b** multiply by 100%
$$\frac{7}{20} = \frac{35}{100} = 35\%$$ (×5)
$$\frac{7}{20} \times 100\% = 35\%$$

key 7 ÷ 20 × 100 =

✓ All fractions convert to either a **terminating** or **recurring decimal**.

Examples $\frac{3}{5} = 0.6$ ← terminating decimal

$\frac{4}{9} = 0.444444444 ...$ ← recurring decimal
$= 0.\dot{4}$ The dot above the 4 shows that the digit repeats.

✓ All recurring decimals are **exact fractions**.
You should know these.

$0.333333333 ... = \frac{1}{3} (\frac{3}{9})$ $\qquad$ $0.666666666 ... = \frac{2}{3} (\frac{6}{9})$ $\qquad$ $0.111111111 ... = \frac{1}{9}$ $\qquad$ $0.999999999 = \frac{9}{9} = 1$

(A) Quick Questions

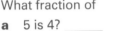 Write fractions in their simplest form.

What fraction of

a 5 is 4? _____ $\qquad$ **b** 20 is 15? _____ $\qquad$ **c** 240 is 160? _____ $\qquad$ **d** 6 is 8? _____

e 10 cm is 10 mm? _____ $\qquad$ **f** 10 cm is 2 mm? _____ $\qquad$ ***g** 10 kg is 400 g? _____

Write these as fractions.

h 0.666666666 ... _____ $\qquad$ **i** 0.555555555 ... _____ $\qquad$ **j** 0.999999999 ... _____

(B) The Armitage

a A large new hotel purchased these beds.
80 single beds $\qquad$ 150 double beds $\qquad$ 170 queen-sized beds
What percentage of the beds were **i** single _____ **ii** queen-sized? _____

b 190 of the 250 rooms were full on the first night. What proportion is this as a decimal? _____

c The hotel's restaurant cooked 28 kg of chicken on their opening night.
They had 21 kg of chicken still uncooked at the end of the night.
What fraction of the chicken was cooked? _____

(C) May Day Regatta

Note: The times were put into class intervals 100 ≤ t < 105 etc.

This frequency diagram shows the times taken by yachts in a race.

a What fraction of the yachts took
i less than 110 mins? _____ $\qquad$ **ii** 115 mins or more? _____

b What percentage of yachts took 110 mins or more, but less than 120 mins? _____

Yachting race times

(frequency diagram: Frequency axis 0–10, Minutes axis 100–125)

(D) Shady shapes

Find the percentage of each shape that is shaded.

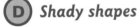 All of the shapes are drawn on squared paper.

a _____ $\qquad$ **b** _____ $\qquad$ **c** _____ $\qquad$ **d** _____

How did you find this? **EASY** **OK** **HARD**

20 Comparing proportions

Let's look at ...
● comparing fractions, decimals and percentages

These are the points you need to know.

✓ We can **compare proportions** using fractions, decimals or percentages.

✓ When comparing fractions we either
  1 convert them to decimals **or**
  2 write them with a common denominator.

Example Compare $\frac{3}{4}$ and $\frac{4}{7}$.

$\frac{3}{4} = \frac{21}{28}$ $\frac{4}{7} = \frac{16}{28}$

28 is the LCM of 4 and 7.   or

$\frac{21}{28} > \frac{16}{28}$

so $\frac{3}{4} > \frac{4}{7}$

$\frac{3}{4} = 0.75$

$\frac{4}{7} = 0.56$ (2 d.p.)

$0.75 > 0.56$

so $\frac{3}{4} > \frac{4}{7}$

Ⓐ *Order it*

Write > or < in the box to make these correct.

a $\frac{1}{5} \square \frac{1}{4}$ **b** $\frac{2}{3} \square \frac{7}{12}$ **c** $\frac{4}{7} \square \frac{3}{5}$ **d** $60\% \square \frac{16}{25}$ **e** $55\% \square \frac{5}{9}$

Put these in order from smallest to largest.

f $\frac{3}{4}, \frac{2}{3}, \frac{5}{8}, \frac{1}{2}$ _____

g $0.68, \frac{3}{5}, 55\%, \frac{13}{20}$ _____

h $\frac{1}{3}, 0.3, \frac{4}{9}, 44\%$ _____

i $\frac{7}{25}, 0.31, 29\%, \frac{9}{30}$ _____

Ⓑ *Election time*

Five boys were nominated to be head boy at Avonside College. This table shows the number of votes each received.

a Which boy received 30% of the teachers' votes?

_____

b Which boy received 25% of the students' votes?

_____

c Julia said 'Jacques was equally popular with the teachers and the students.'
Is Julia correct? _____ Explain why or why not. _____
_____

d Which boy **was** equally popular with the teachers and the students? _____

Nominee	Number of teacher votes	Number of student votes
Guy	0	16
Jacques	4	4
Manzoor	2	40
Ross	3	90
Gareth	1	50
Total	**10**	**200**

Ⓒ *Inside or out?*

In a biology experiment, Eleanor planted some seeds inside and some outside. This table shows how many germinated.

	Inside	Outside
Seeds planted	213	257
Seeds germinated	162	186

a For the inside seeds, calculate the number of seeds which germinated as a percentage of the number of seeds planted. Give your answer to 1 d.p. ____

b Show that the proportion of seeds which germinated was greater for the inside plants than the outside plants.

---

21 Adding and subtracting fractions

 except

Let's look at ...
- adding and subtracting fractions with different denominators
- adding and subtracting mixed numbers

These are the points you need to know.

✓ To **add and subtract fractions with different denominators** we use equivalent fractions. We find the LCM of the denominators then write both fractions with this as the denominator.

Examples $\frac{3}{8} + \frac{2}{3} = \frac{9}{24} + \frac{16}{24}$

$= \frac{9+16}{24}$

$= \frac{25}{24}$

$= 1\frac{1}{24}$

$\frac{3}{8} = \frac{9}{24}$ and $\frac{2}{3} = \frac{16}{24}$

$\frac{5}{8} - \frac{2}{5} = \frac{25}{40} - \frac{16}{40}$

$= \frac{9}{40}$

✓ To **add or subtract mixed numbers**, we write each as an improper fraction first.

Examples ● $2\frac{3}{4} + 1\frac{1}{3} = \frac{11}{4} + \frac{4}{3}$

$= \frac{33}{12} + \frac{16}{12}$

$= \frac{49}{12}$

$= 4\frac{1}{12}$

or $2\frac{3}{4} + 1\frac{1}{3} = 2 + 1 + \frac{3}{4} + \frac{1}{3}$

$= 3 + \frac{9}{12} + \frac{4}{12}$

$= 3 + \frac{13}{12}$

$= 3 + 1\frac{1}{12} = 4\frac{1}{12}$

When adding, we can add the whole numbers first.

● $4\frac{1}{5} - 2\frac{1}{2} = \frac{21}{5} - \frac{5}{2}$

$= \frac{42}{10} - \frac{25}{10}$

$= \frac{17}{10}$

$= 1\frac{7}{10}$

✓ We can add and subtract fractions using the $a^{b/c}$ **key** on a calculator.

Example To find $2\frac{3}{8} + 3\frac{1}{3}$ key $\boxed{2}\,\boxed{a^{b/c}}\,\boxed{3}\,\boxed{a^{b/c}}\,\boxed{8}\,\boxed{+}\,\boxed{3}\,\boxed{a^{b/c}}\,\boxed{1}\,\boxed{a^{b/c}}\,\boxed{3}\,\boxed{=}$

to get $\boxed{5\,⌐17⌐24}$. We read this as $5\frac{17}{24}$.

Ⓐ Getting started

You will need some working paper for this question.

 Write all of your answers in their simplest form.

a Complete these number chains.

i 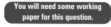 $\boxed{\frac{1}{5}}\xrightarrow{+\frac{2}{5}}\boxed{\frac{3}{5}}\xrightarrow{+\frac{4}{5}}\boxed{}\xrightarrow{-\frac{3}{5}}\boxed{}\xrightarrow{+1\frac{1}{5}}\boxed{}\xrightarrow{-1\frac{5}{6}}\boxed{}\xrightarrow{+\frac{1}{3}}\boxed{}\xrightarrow{+\frac{3}{8}}\boxed{}$

ii $\boxed{1\frac{1}{4}}\xrightarrow{+2\frac{1}{2}}\boxed{}\xrightarrow{+1\frac{1}{8}}\boxed{}\xrightarrow{-3\frac{1}{4}}\boxed{}$ **iii** $\boxed{2\frac{2}{3}}\xrightarrow{+1\frac{1}{2}}\boxed{}\xrightarrow{-2\frac{1}{3}}\boxed{}\xrightarrow{+4\frac{3}{5}}\boxed{}$

b Calculate these and show the answers on the number line.

i $\frac{2}{3} + \frac{5}{6}$ **ii** $\frac{9}{24} + \frac{7}{8}$

0   1   2

c Use your calculator to find the answers to these.

i $\frac{7}{15} + \frac{2}{7}$ _____ **ii** $\frac{8}{11} + \frac{4}{9} - \frac{3}{5}$ _____ **iii** $4\frac{1}{5} - 2\frac{3}{13}$ _____ **iv** $6\frac{2}{3} + 1\frac{5}{12} - 4\frac{3}{7}$ _____

Ⓑ Jo's kitchen

a Jo has $\frac{1}{2}$ kg of butter. She uses $\frac{1}{6}$ kg in her baking. How much butter is left? ____

b Jo has 4 m of tin foil. She lines the top tray in her oven with $1\frac{1}{3}$ m of foil, and the bottom tray with $1\frac{4}{5}$ m. How much is left? ____

Ⓒ Puzzles

a Write a number in each square to make these true. **i** $\frac{1}{2} + \boxed{} = \frac{5}{6}$ **ii** $\frac{1}{5} + \boxed{} = \frac{11}{30}$ **iii** $\boxed{} + \boxed{} = \frac{13}{42}$

b Write the next number in these sequences. **i** $\frac{1}{6}, \frac{5}{6}, 1\frac{1}{2}, 2\frac{1}{6}, 2\frac{5}{6}$, _____ **ii** $\frac{1}{2}, 1\frac{7}{8}, 3\frac{1}{4}, 4\frac{5}{8}$, _____

c Complete these magic squares.

i
$\frac{1}{4}$		
$\frac{7}{8}$		$\frac{1}{8}$
$\frac{3}{8}$		

ii
$1\frac{1}{6}$	$2\frac{5}{6}$	
	$2\frac{1}{2}$	
	$2\frac{1}{6}$	

iii
$1\frac{1}{10}$		
	$2\frac{1}{2}$	
$1\frac{4}{5}$	$3\frac{9}{10}$	

 In a magic square the numbers in every row, column and diagonal add to the same total.

How did you find this? **EASY** **OK** **HARD**

22 Fraction and percentage of

 except **B** b

Let's look at ...
● finding fractions and percentages of quantities mentally
● finding fractions and percentages of quantities using a calculator

---

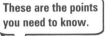 These are the points you need to know.

✓ We can find **fractions or percentages of quantities**.
Try to use a mental method first.

Examples $\frac{3}{5}$ of 45   35% of 80
$\frac{1}{5}$ of 45 = 9   10% of 80 = 8
$\frac{3}{5}$ of 45 = 3 × 9   30% of 80 = 24
        = **27**   5% of 80 = 4
                    35% of 80 = 24 + 4 = **28**

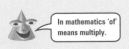 In mathematics 'of' means multiply.

✓ For harder calculations we sometimes **use a calculator**.

Example   To find $17\frac{1}{2}$% of 125

either   key 17·5 ÷ 100 × 125 = to get 21·875   **using fractions**
or   key 0·175 × 125 = to get 21·875   **using decimals**
or   key 17·5 × 1·25 = to get 21·875   **multiplying by 1%**

(A) Did you know?

			S									S			
70	10	40	6	72	10	72	18	72		22	40	6	40	18	70

						S							
40	12	210	72	54	22	6		75	96		72	75	36	39	70

36	54	40		110	147		210	40	18		147	36	54	70	10

Write the letter beside each question above its answer in the box. Find the answers mentally.

S $\frac{1}{4}$ of 24 = 6   **R** $\frac{3}{4}$ of 24   **D** $\frac{2}{5}$ of 55   **X** $\frac{3}{7}$ of 28   **H** 20% of 50

N 60% of 90   **B** $\frac{5}{6}$ of 90   **O** $\frac{3}{8}$ of 96   **T** 25% of 280   **K** 55% of 200

M 49% of 300   **E** $1\frac{1}{4}$ of 32   **U** 65% of 60   **A** 45% of 160   **P** $1\frac{3}{4}$ of 120

Y $2\frac{2}{5}$ of 40

(B) Which is greater?

Circle the greater number in each pair.

a Answer these mentally.   **i** 75% of 24 $\frac{3}{5}$ of 35   **ii** $\frac{4}{9}$ of 81 $\frac{5}{7}$ of 42   **iii** 30% of 210 $\frac{2}{5}$ of 150

 b Use your calculator to answer these.
i $\frac{2}{3}$ of 68   84% of 56   **ii** 6% of £36·50 $17\frac{1}{2}$% of £13   **iii** 58·3% of 2·1 kg $66\frac{2}{3}$% of 1·75 kg

(C) What's missing?

Fill in the gaps.   **a** $\frac{1}{2}$ of 40 = $\frac{1}{4}$ of ____   **b** $\frac{3}{4}$ of 80 = $\frac{1}{2}$ of ____   **c** $\frac{1}{3}$ of 90 = $\frac{2}{3}$ of ____

(D) Wordy wonderings

a A school has 200 students. 25% are Scottish, $\frac{2}{5}$ are English and 30 are Indian. The rest are Welsh. How many students are Welsh? ____

*b Danielle spent $\frac{2}{3}$ of her birthday money on a new pair of jeans. She spent $\frac{1}{4}$ of what was left on lunch. She then had £15 still to spend.
How much birthday money did she get? ____

---

23 Multiplying fractions

Let's look at ...
● multiplying fractions and mixed numbers

✓ When **multiplying fractions** we
1 write whole numbers or mixed numbers as improper fractions
2 cancel if possible
3 multiply the numerators
multiply the denominators.

Example $1\frac{1}{2} \times 2\frac{2}{3} = \frac{3}{2}^1 \times \frac{8}{3}^4$
$= \frac{4}{1}$
$= 4$

These are the points you need to know.

You can cancel any numerator with any denominator.

(A) *Pyramids*

Multiply two fractions to work out the number that goes on top of them.

a

$\frac{3}{4} \times \frac{1}{2} = \frac{3}{8}$

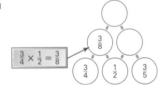

b

c

(B) *Turkish rugs*

Calculate the area (A) of each of these floor rugs.

a $2\frac{2}{3}$ m

$1\frac{1}{2}$ m

$A = $ _____

b $2\frac{2}{5}$ m

$3\frac{1}{3}$ m

$A = $ _____

c $3\frac{3}{4}$ m

$1\frac{1}{10}$ m

$A = $ _____

d $1\frac{5}{6}$ m

$1\frac{5}{6}$ m

$A = $ _____

(C) *True or false?*

Decide whether each of these is true (T) or false (F).

Remember: In maths, 'of' means multiply.

a   One-half of two-thirds is one-third.   ____

b   Four-fifths of three-eighths is three-fifths.   ____

c   One-third of three-quarters is the same as three-quarters of one-third.   ____

d $\frac{9}{11} \times \frac{6}{7} = \frac{3}{11} \times \frac{2}{7}$ _____

e $\left(\frac{4}{3}\right)^2 = \frac{4^2}{3^2}$ _____

f $\left(1\frac{1}{2}\right)^2 = 1\frac{3}{4}$ _____

g $\frac{3}{4} \times \frac{16}{21} \times \frac{14}{15} = \frac{8}{9}$ _____

h $\frac{5}{36} \times 24 \times 12 = 40$ _____

i $\frac{1}{5}\left(2 - \frac{3}{4}\right) = \frac{1}{4}$ _____

(D) *Your turn*

a   Find three different pairs of fractions which multiply to these.

i $\frac{9}{16}$ _____ , _____ , _____

ii $\frac{14}{15}$ _____ , _____ , _____

*b   Use three of these five fractions to make these true.

$\frac{3}{4}, \frac{4}{5}, \frac{5}{6}, \frac{6}{7}, \frac{7}{8}$

i $\boxed{} \times \boxed{} \times \boxed{} = \frac{1}{2}$

ii $\boxed{} \times \boxed{} \times \boxed{} = \frac{3}{5}$

How did you find this?   EASY   OK   HARD

24 Dividing fractions

Let's look at ...
● using the inverse rule to divide by fractions

These are the points you need to know.

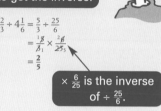

✓ When we **divide fractions** we use the **inverse rule**. To divide by a fraction, multiply by the inverse.

Turn the fraction you are dividing by upside down to get the inverse.

Examples $\frac{2}{3} \div \frac{5}{9} = \frac{2}{3_1} \times \frac{9^3}{5}$
$= \frac{6}{5}$
$= 1\frac{1}{5}$

$\times \frac{9}{5}$ is the inverse of $\div \frac{5}{9}$.

$1\frac{3}{4} \div \frac{5}{8} = \frac{7}{4_1} \times \frac{8^2}{5}$
$= \frac{14}{5}$
$= 2\frac{4}{5}$

$\times \frac{8}{5}$ is the inverse of $\div \frac{5}{8}$.

$1\frac{2}{3} \div 4\frac{1}{6} = \frac{5}{3} \div \frac{25}{6}$
$= \frac{5}{3_1} \times \frac{6}{25_5}$
$= \frac{2}{5}$

$\times \frac{6}{25}$ is the inverse of $\div \frac{25}{6}$.

(A) Put on your shades

Answer each question then shade the answer in the grid.

The shaded squares will form a number. This number is the length, in years, of summer on Uranus.

HELP

Need help with multiplying fractions? See page 24.

a $\frac{2}{5} \div 2$ ____ **b** $4 \div \frac{2}{9}$ ____ **c** $\frac{3}{7} \div 6$ ____

d $15 \div \frac{3}{5}$ ____ **e** $5 \div \frac{3}{4}$ ____ **f** $6 \div \frac{2}{9}$ ____

g $\frac{2}{3} \div \frac{1}{3}$ ____ **h** $\frac{5}{6} \div \frac{3}{4}$ ____ **i** $\frac{3}{10} \div \frac{5}{8}$ ____

j $2\frac{1}{2} \div \frac{2}{5}$ ____ **k** $1\frac{3}{5} \div \frac{1}{4}$ ____ **l** $2\frac{3}{4} \div \frac{2}{3}$ ____

m $4\frac{1}{6} \div \frac{10}{11}$ ____ **n** $1\frac{2}{7} \div 1\frac{5}{7}$ ____ **o** $2\frac{2}{5} \div 5\frac{1}{4}$ ____

*p $\dfrac{1-\frac{1}{4}}{1-\frac{5}{6}}$ ____

$1\frac{5}{6}$	27	$4\frac{1}{8}$	$1\frac{1}{9}$	$2\frac{1}{2}$	$4\frac{1}{2}$
$\frac{3}{16}$	$\frac{4}{5}$	$\frac{5}{8}$	18	$\frac{2}{9}$	$\frac{1}{5}$
$\frac{2}{5}$	$\frac{12}{25}$	$\frac{3}{4}$	$4\frac{7}{12}$	$\frac{10}{33}$	$6\frac{2}{5}$
5	$6\frac{2}{3}$	$\frac{9}{8}$	$2\frac{1}{4}$	$\frac{8}{9}$	25
$\frac{18}{7}$	$6\frac{1}{4}$	$\frac{16}{35}$	$\frac{1}{14}$	$\frac{4}{7}$	2
$\frac{3}{42}$	$1\frac{1}{3}$	15	1	$1\frac{4}{7}$	$\frac{12}{9}$

Summer on Uranus lasts for ____ years.

(B) Show your working

Calculate these. Show your working.

a $\frac{4}{5} \div \frac{6}{15}$ **b** $3\frac{1}{4} \div \frac{7}{8}$ **c** $1\frac{3}{5} \div 5\frac{1}{3}$

(C) Three problems

a How many tenths are in $5\frac{1}{5}$? _____

b Pip uses $\frac{3}{4}$ m of ribbon for each parcel she ties. How many parcels can she tie with 6 m of ribbon? _____

c If $2\frac{1}{2}$ kg of turkey costs £8, what is the cost per kg? _____

(D) Your turn

a Find three different pairs of fractions which give the answer $\frac{3}{8}$ when divided. ____, ____, ____

b Complete this equation in three different ways.

$\frac{4}{\square} \div \frac{\square}{4} = \frac{8}{9}$ $\frac{4}{\square} \div \frac{\square}{4} = \frac{8}{9}$ $\frac{4}{\square} \div \frac{\square}{4} = \frac{8}{9}$

How did you find this? EASY OK HARD

25 Percentage change

Let's look at ...
● calculating percentage increases and decreases
● giving an increase or decrease as a percentage
● calculating the original amount after a percentage change

✓ We can calculate a **percentage increase or decrease**.
To **increase an amount** by 35% we multiply by 135% or 1·35 (100% + 35%).
To **decrease an amount** by 20% we multiply by 80% or 0·8 (100% − 20%).

Some answers need to be rounded sensibly.

✓ We can give an **increase or decrease as a percentage increase or decrease**.
% increase = $\frac{\text{actual increase}}{\text{original amount}} \times 100\%$ % decrease = $\frac{\text{actual decrease}}{\text{original amount}} \times 100\%$

These are the points you need to know.

✓ We can find the **original amount** if we know the amount after the percentage change.
We use **inverse operations** or the **unitary method** or **algebra**.

Example   After a 25% increase, a box of chocolates costs £3·40.
We can find the original price in one of these ways.

Unitary method	**Inverse operations**	**Algebra**
Find 1% first.		$1\cdot25 \times c = £3\cdot40$ where c represents original price of chocolates
£3·40 represents 125%.		
£3·40 ÷ 125 represents 1%.		$c = \frac{£3\cdot40}{1\cdot25}$
£3·40 ÷ 125 × 100 represents 100%.		$= £2\cdot72$
£3·40 ÷ 125 × 100 = **£2·72**		

Inverse operations:
original price of chocolates × 1·25 → new price £3·40   + 1·25 ←

Original price of chocolates × 1·25 = £3·40
Original price of chocolates = $\frac{£3\cdot40}{1\cdot25}$ = **£2·72**

(A) *On the High Street*

Hint i is an increase and ii is a decrease.

a A supermarket buys tubs of ice cream at £3 each.

  i It prices the ice cream to make a 40% profit. How much does each tub cost? _____

  ii When the ice cream gets near its 'best before' date, the price is discounted by 25%. How much does the ice cream cost then? _____

b A bicycle depreciated in value from £245 to £180. What percentage decrease is this? _____

Hint See the second tick in the box above.

APPLE CRISPS NEW
Now contain 15% more!

c This packet of apple crisps contains 368 g of crisps. What mass of crisps was in the old style packets? _____

Hint See the third tick in the box above.

(B) *Population boom?*   *Round sensibly.*

a Sarah's village had a population of 150 a year ago. Since then it has increased by 8%. What is the population now? _____

b This table gives the populations of two towns.

  i The population of Gordonville increased by 20·18% from 1961 to 1991. Complete the table.

Year	1931	1961	1991
Dunsten	2981	3743	6250
Gordonville	634	9410	

  ii Find the percentage increase in Dunsten between
    ● 1931 and 1961 _____   ● 1961 and 1991 _____

c The number of babies born at a hospital this year was 19% up on last year. This year 695 babies were born. How many were born last year? _____

(C) *Take your pick*

Here are four calculations.   **A** 45 × 1·7   **B** 45 × 0·07   **C** 45 × 0·7   **D** 45 × 1·07

a Which calculation answers this question: '**What is 45 increased by 7%?**' _____

b Choose two of the other calculations. For each write a question **about percentages** that this calculation represents.

How did you find this? [**EASY**] [**OK**] [**HARD**]

26 Proportionality

Let's look at ...
● using proportional reasoning to solve problems

These are the points you need to know.

✓ **Proportional reasoning** can be used to solve problems.
The **unitary method** or **constant multiplier method** can be used.

Example If 8 muffins cost £6·75 we can find the cost of 12 muffins like this.

Unitary method: We find the cost of **one** first.
Cost of 1 muffin = $\frac{£6·75}{8}$
Cost of 12 muffins = $12 \times \frac{£6·75}{8}$ = **£10·13 (nearest penny)**

Constant multiplier method
Ratio of number of muffins required to number given = 12 : 8
The constant multiplier is $\frac{12}{8}$.
Cost of 12 muffins = $\frac{12}{8} \times £6·75$
    = **£10·13 (nearest penny)**

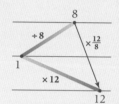

(A) Sports talk

a It costs £7 to hire a squash court for 2 hours. How much would 6 hours cost? _____

b Four golf balls cost £10. How much would 12 golf balls cost? _____

c Matthew bowls four cricket overs in 11 minutes. How long does he take to bowl ten overs, if he always bowls at the same rate? _____

d Ashleigh pays £43 for five basketballs. How much will she pay for eight basketballs? _____

(B) Tuna for dinner

Adapt this recipe to feed 10 people.

Tuna pasta (for 4 people)	
• 50 g butter	• 500 g pasta
• 2 tbsp flour	• 190 g tuna
• 380 ml milk	• 250 g grated cheese

Tuna pasta (for 10 people)	
•	•
•	•
•	•

(C) What's for lunch?

Harriet is thinking about what to have for lunch. She looks at labels on a tin of baked beans and a tin of spaghetti.

a How much protein would 250 g of spaghetti provide?
_____

Spaghetti each 150 g provides	
Energy	405 kJ
Protein	3·6 g
Fat	0·2 g
Carbohydrate	18·9 g

Baked beans each 125 g provides	
Energy	550 kJ
Protein	4·5 g
Fat	0·1 g
Carbohydrate	15·5 g

b Each tin contains 420 g of food. If Harriet eats a whole tin, which food would provide more carbohydrate? _____ How much more? _____

*(D) Summer holiday

a Use **£1 = 1·46 Euros** to work out how much **65p** is in Euros. Show your working.

65p = _____ Euros

b Use **1·46 Euros = £1** to work out how much **7·3 Euros** is in pounds. Show your working.

7·3 Euros = £_____

c Use **£1 = 1·46 Euros** and **£1 = 10·5 Danish Krona** to work out how much **1 Euro** is in **Danish Krona**. Show your working and round sensibly.

1 Euro = _____ Danish Krona

How did you find this? [EASY] [OK] [HARD]

27 Ratio

 except

Let's look at ...
- simplifying ratios
- comparing ratios

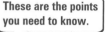
These are the points you need to know.

✓ We **simplify ratios by cancelling**. A ratio in its simplest form does not have fractions or decimals.

Examples

$+4 \left(\overset{12 : 16}{} \right) +4$
$= 3 : 4$

$\times 3 \left(\overset{1\frac{2}{3} : 2}{} \right) \times 3$
$= 5 : 6$

✓ To **compare ratios** we write each in the form $1 : m$ or $m : 1$.

Example The ratios of butter to sugar in recipe A and recipe B are 2 : 3 and 7 : 11.
To compare these we write both in the form $1 : m$.

Recipe A
$+2 \left(\overset{2 : 3}{} \right) +2$
$= 1 : 1 \cdot 5$

Recipe B
$+7 \left(\overset{7 : 11}{} \right) +7$
$= 1 : 1 \cdot 6 \text{ (1 d.p.)}$

We divide both parts by the smaller number in the ratio.

There is a greater proportion of butter in recipe A because $\frac{1}{1 \cdot 5} > \frac{1}{1 \cdot 6}$.

(A) Quick questions

Write each ratio in its simplest form.

a 10 : 15 ____ **b** 42 : 49 ____ **c** 9 : 18 : 21 ____ **d** 2·5 : 7 ____

e $4 : 2\frac{1}{2}$ ____ **f** $1\frac{1}{4} : 5$ ____ **g** 25 cm : 1 m ____ **h** 1 hour : 40 mins ____

Write these ratios in the form $1 : m$. Round to 2 d.p. if necessary.

i 4 : 11 ____ **j** 5 : 63 ____ **k** 7 : 38 ____ **l** 29 : 36 ____

(B) Romeo and Juliet

Romeo is 18 years old. Juliet is exactly **8 years younger**, so this year she is 10 years old.

This year, the ratio of Romeo's age to Juliet's age is 18 : 10.

18 : 10 written as simply as possible is **9 : 5**.

a When Romeo is **28**, what will be the ratio of Romeo's age to Juliet's age?
Write the ratio as simply as possible. _____ : _____

b When Juliet is **32**, what will be the ratio of Romeo's age to Juliet's age?
Write the ratio as simply as possible. _____ : _____

(C) Wordy wonderings

a Isabella pours 500 mℓ of apple juice and 1·5 ℓ of orange juice into a jug.
Write the ratio of apple to orange juice as a ratio in its simplest form. _____

*b The ratio *nitrogen (N) : potassium (K)* in two fertilisers is given.

Fertiliser A	Fertiliser B
N : K	N : K
12 : 5	9 : 4

  i Change each ratio to the form $m : 1$.
      Fertiliser A _____   Fertiliser B _____
  ii Which fertiliser has the greater proportion of nitrogen? _____

*c The ratio of males to females at two football clubs is given.
   Avon United 82 : 15   **Heathcote United** 43 : 8
   By changing each ratio to the form $m : 1$, say which team has the greater proportion of females.
   Explain. _____

   _____

How did you find this? 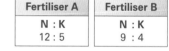 EASY   OK   HARD

28 Ratio and proportion problems

Let's look at ...
● solving ratio and proportion problems

These are the points you need to know.

✓ We can solve **ratio and proportion** problems.
Ratio compares part to part. Proportion compares part to whole.

Example In a photo a 1·8 m tall man is 3 cm tall.
In the same photo, a child is 2·1 cm tall.
How tall is the child in real life?

Answer
In real life the man is 1·8 m or 180 cm tall and in the photo he is 3 cm.

height in photo → ⟦ ? ⟧ → height in real life

3 cm → ⟦ ×$\frac{180}{3}$ ⟧ → 180 cm

2·1 cm → ⟦ ×$\frac{180}{3}$ ⟧ → 126 cm   The child is **1·26 m** tall in real life.

A Ma's bakery

A bakery has these pies for sale.

6 mince   18 steak   12 chicken   4 potato top

a Write each of these as a ratio in its simplest form.

i mince : chicken   _____   **ii** steak : potato top   _____

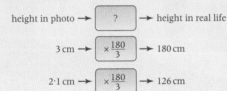

Give each answer in b as a fraction, decimal and percentage.

b What proportion of the pies are

i potato top?   _____, _____, _____   **ii** steak?   _____, _____, _____

B Orange please

Cara has two brands of orange juice concentrate, Zesty and Bounce.

Zesty — Mix 1 part Zesty with 5 parts water

Bounce — Mix 1 part Bounce with 8 parts water

a How much water should Cara mix with 60 mℓ of Zesty concentrate? _____

b How much Bounce concentrate should Cara mix with 720 mℓ of water? _____

c Cara thinks that after mixing, the Zesty drink is 20% concentrate. Is she right? _____
Explain. _____

C More problems

a A 750 m² plot is divided into house and garden in the ratio 7 : 8.
What is the area of the garden? _____

b The angles in a triangle are in the ratio 3 : 5 : 4. Find the sizes of the three angles. ____, ____, ____

c A school requires that the ratio of adults to students on an outing be 1 : 6.
If 228 students will be on the outing, how many adults are required? _____

d A geologist found that the ratio of sedimentary to igneous rocks in a collection was 7 : 6.
If 84 of the rocks were sedimentary, how many were igneous? _____

e An artist sketches scale drawings of stage sets. In her sketch a table is 8 cm high. It will be built 120 cm high. Her sketch has a 28 cm long boat. How long should the boat be when built? _____

*f William had two science tests on forces. He scored 25% better in the second one than in the first one.
What is the ratio of his score in the first test to his score in the second test? _____

How did you find this?   ⟦ EASY ⟧   ⟦ OK ⟧   ⟦ HARD ⟧

29 Starting algebra

Let's look at ...
● equations, formulae, functions and identities

These are the points you need to know.

✓ Here are some examples of **equations**.

$$2x - 3 = 7 \qquad \frac{5p - 3}{2} = 6 \qquad 4(x + 2) = 24 \qquad 5x + 3 = 2x + 9$$

Each has an equals sign. The unknown has a particular value.

✓ A **formula** gives the relationship between variables.

$s = \frac{D}{T}$ (s = speed, D = distance, T = time)   The variables stand for something specific.

✓ A **function** is a special sort of equation. It gives the relationship between two variables, usually x and y. $y = 2x - 3$ is a function. If we know the value of x we can find the value of y.

$x \rightarrow \boxed{\text{multiply by 2}} \rightarrow \boxed{\text{subtract 3}} \rightarrow y$

✓ An **identity** is true for **all** values of the unknown. For all values of x, the left-hand side will always equal the right-hand side.

Example $4(x + 2) \equiv 4x + 8$ Choose $x = 3$

$$4(x + 2) = 4(\mathbf{3} + 2) \qquad \text{and} \qquad 4x + 8 = 4 \times \mathbf{3} + 8$$
$$= 4 \times 5 \qquad\qquad\qquad\qquad = 12 + 8$$
$$= 20 \qquad\qquad\qquad\qquad\quad = 20$$

left-hand = right-hand side

$\equiv$ means is identically equal to.

A Sort them out

Copy each statement from the box into the correct rectangle.

● $y = 3x + 2$

● $P = I^2 R$ where P is power in watts, I is current in amps and R is resistance in ohms

● $y = 5x - 1$

● $4x + 5 = 10$

● $V = \frac{1}{3} Ah$ where V is volume of cone in m³, A is area of base in m² and h is height in m

● $3(x - 2) = 12$

● $a = \frac{d}{t^2}$ where a is acceleration in m/s², d is distance in m and t is time in seconds

● $\frac{x - 5}{3} = 6$

● $y = \frac{x + 6}{4}$

● $2x + 1 = 3x - 4$

● $y = 20 - \frac{x}{3}$

Equations

Formulae

Functions

B Check them

a   Show that $6(x + 2) \equiv 6x + 12$ is true for these values of x.

i $x = 1$

ii $x = 8$

iii $x = {}^-2$

b   Check if these statements are identities by substituting values for x. You could try $x = 0{\cdot}3$ and $x = {}^-2$.

i $3(x + 4) \equiv 3x + 12$ YES/NO

ii $6x - 12 \equiv 6(x - 2)$ YES/NO

Multiply out the brackets to check.

C True or false?

Write true or false for each of these and explain your choice.

a $4x + 1 = 9$ is a function. _____ because _____

_____

b $F = ma$ where F is force in Newtons, m is mass in g and a is acceleration in m/s²   is a formula. _____ because _____

_____

c $5(2x + 1) \equiv 5x + 5$ is an identity. _____ because _____

_____

How did you find this?   EASY   OK   HARD

30 Inequalities

Let's look at ...
- the meaning of $<$, $>$, $\leqslant$ and $\geqslant$
- the rules of working with inequalities

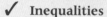

These are the points you need to know.

✓ **Inequalities**

$<$ means 'is less than'. $\qquad$ $>$ means 'is greater than'.

$\leqslant$ means 'is less than or equal to'. $\qquad$ $\geqslant$ means 'is greater than or equal to'.

Examples $\quad x \leqslant 7 \qquad\qquad 2 \leqslant y < 4$

$\qquad\qquad\quad$ x is less than $\qquad$ y is greater than or equal
$\qquad\qquad\quad$ or equal to 7 $\qquad$ to 2 but less than 4

✓ If we start with an inequality, this will stay true if we
- add or subtract the same number to both sides
- multiply or divide both sides by the same **positive** number.

✓ If we multiply or divide both sides by the same **negative** number, the inequality sign must be **reversed**.

Examples $\quad ^-2 > ^-5 \qquad\qquad\qquad ^-4 < 2$
$\qquad\qquad ^-2 - 3 > ^-5 - 3 \qquad\qquad ^-4 \times ^-3 > 2 \times ^-3 \qquad$ multiply by a negative number
$\qquad\qquad ^-5 > ^-8 \qquad\qquad\qquad 12 > ^-6 \qquad\qquad\qquad$ inequality sign reversed

(A) Pick the best

Circle the inequality which best describes the statement.

a Aaron's cat weighs more than 7 kg.

$\qquad c < 7 \qquad c > 7 \qquad c \leqslant 7 \qquad c \geqslant 7$

b It takes Dean at least 25 minutes to get to school.

$\qquad t < 25 \qquad t > 25 \qquad t \leqslant 25 \qquad t \geqslant 25$

c The maximum tyre pressure for my car is 28 lb/sq. in.

$\qquad p < 28 \qquad p > 28 \qquad p \leqslant 28 \qquad p \geqslant 28$

d Ruth's team have never scored more than 9 goals in a football game.

$\qquad g \geqslant 9 \qquad g < 9 \qquad g \leqslant 9 \qquad g > 9$

e Mark's sunflowers all grew to at least 0·5 m, but none reached 3 m.

$\qquad 0{\cdot}5 < h \leqslant 3 \qquad 0{\cdot}5 \leqslant h < 3 \qquad 0{\cdot}5 < h < 3 \qquad 0{\cdot}5 \leqslant h \leqslant 3$

(B) Find n

Use the clues provided to find all possible values for n in each question.

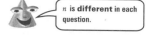 n is **different** in each question.

a
- n is positive
- n is odd
- $n < 11$

_____

b
- n is positive
- n is even
- $n \leqslant 6$

_____

c
- n is a square number
- $4 < n \leqslant 49$

_____

d
- n is a negative whole number
- $^-5 < n$

_____

(C) Yes or no?

Dale started with this inequality: $\boxed{^-4 < 2}$

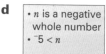 Write **yes** or **no** for each answer in **a** and **b**.

a He added 4 to both sides.
$\quad$ Is the inequality still true? _____ $\qquad$ $^-4 + \mathbf{4} < 2 + \mathbf{4}$

b Check to see if the inequality is still true if you do these to both sides:

$\quad$ **i** $\;$ add 8 _____ $\qquad$ **ii** $\;$ subtract 1 _____ $\qquad$ **iii** $\;$ subtract $6\frac{1}{2}$ _____ $\qquad$ **iv** $\;$ multiply by 3 _____

$\quad$ **iv** $\;$ divide by 2 _____ $\qquad$ **v** $\;$ multiply by $^-4$ _____ $\qquad$ **vi** $\;$ divide by $^-2$ _____

c You should have answered '**no**' to two questions in part **b**.

$\quad$ What do you need to do to make these inequalities correct? _____

$\qquad\qquad\qquad$ How did you find this? $\quad$ **EASY** $\quad$ **OK** $\quad$ **HARD**

31 Solving equations

Let's look at ...
● **writing and solving equations**

These are the points you need to know.

✓ When we **write an equation** we must choose a suitable unknown.

✓ We can **solve equations** using **inverse operations** or by **transforming both sides**.

Example $3(n + 2) + 4 = 25$	Inverse operations	Transforming both sides
	$3(n + 2) + 4 = 25$	$3(n + 2) + 4 = 25$
	$3(n + 2) = 25 - 4$	$3n + 6 + 4 = 25$
	$n + 2 = \frac{21}{3}$	$3n + 10 = 25$
	$n = 7 - 2$	$3n + 10 - 10 = 25 - 10$
	$= 5$	$3n = 15$
		$\frac{3n}{3} = \frac{15}{3}$
		$n = 5$

(A) Listen hard

You will need some extra paper for your working.

	$\overline{3}$ **O**											
$\overline{33}$	$\overline{3}$	$\overline{15}$	$\overline{9}$	$\overline{6}$	$\overline{30}$	$\overline{2}$	$\overline{8}$	$\overline{10}$	$\overline{1}$	$\overline{54}$	$\overline{33}$

| $\overline{7}$ | $\overline{8}$ | $\overline{33}$ | $\overline{30}$ | $\overline{1}$ | $\overline{2}$ | $\overline{30}$ | $\overline{^-6}$ | $\overline{2}$ | $\overline{3}$ **O** | $\overline{15}$ | $\overline{5}$ | $\overline{^-6}$ |

| $\overline{4}$ | $\overline{8}$ | $\overline{30}$ | $\overline{1}$ | $\overline{2}$ | $\overline{30}$ | $\overline{^-6}$ | $\overline{8}$ | $\overline{9}$ | $\overline{8}$ | $\overline{^-5}$ | $\overline{2}$ |

Solve these equations. Write the letter beside each equation above its answer in the box.

O $7x + 1 = 22$ $x = 3$ **R** $4x + 3 = 11$ **D** $6x - 5 = 31$ **V** $26 = 3x - 4$

A $\frac{x}{4} = 2$ **L** $\frac{x}{9} = 6$ **U** $\frac{x}{5} + 4 = 7$ **T** $\frac{x}{6} - 2 = 3$

F $5 = \frac{35}{x}$ **S** $\frac{x+3}{9} = 4$ **N** $\frac{4x-8}{7} = 4$ **E** $3(x + 2) = 9$

G $4(x - 2) + 7 = 19$ **W** $5x + 6 + 3x - 2 = 36$ ***H** $3x - 1 = {}^-19$ ***I** $5x + 6 = {}^-19$

(B) Write carefully

Write and solve equations for these. Show your working.

a
The number in each circle is found by adding the two numbers below it. Find n.

b
The area of this swimming pool is 64 m². Find p.

(C) Look closely

$2g + 3h = 17$ and $4p - 2q = 6$

a Evaluate these.   **i** $4g + 6h$ _____   **ii** $2p - q$ _____

b Use one or both of the equations to write an expression that has a value of 23.

_____ = 23

32 Equations with unknowns on both sides

Let's look at ...
- writing and solving equations with unknowns on both sides

✓ To **solve equations with unknowns on both sides** we 'transform both sides' to get all the unknowns on one side.

Example
$$7p - 3 = 4p + 12$$
$$7p - \mathbf{4p} - 3 = 4p - \mathbf{4p} + 12$$
$$3p - 3 = 12$$
$$3p - 3 + \mathbf{3} = 12 + \mathbf{3}$$
$$3p = 15$$
$$\frac{3p}{3} = \frac{15}{3}$$
$$p = 5$$

 Subtract the smaller number of unknowns from each side.

These are the points you need to know.

(A) Hold on tight!

 You will need some extra paper for your working.

Solve these equations then shade the answers in the grid.
The shaded answers will form a number. This number is the top speed, in miles per hour, of the first train.

a $3n + 3 = n + 7$ $\underline{n = 2}$

b $2 + 7n = 5n + 10$ _____

c $6n + 1 = 4n + 6$ _____

d $2n + 12 = 5n - 9$ _____

e $9n - 6 = 3n + 14$ _____

f $10n + 7 = 15n + 3$ _____

g $6 - 2n = 4n - 3$ _____

h $4(n + 3) = n + 21$ _____

i $2(n + 2) = 4(n - 3)$ _____

j $2n - (n + 1) = 3n + 1$ _____

k $\frac{4(2n + 3)}{14} = 1$ The top speed of the first train was _____ miles per hour.

$\frac{1}{2}$	1	$1\frac{1}{2}$	$^-1$	$2\frac{1}{2}$
$^-5$	$1\frac{1}{3}$	$\frac{1}{4}$	$^-3$	$26\frac{1}{2}$
5	9	2	$3\frac{1}{3}$	7
$2\frac{1}{5}$	6	$2\frac{1}{4}$	10	8
$^-2$	12	4	$\frac{4}{5}$	3

(B) Shapely equations

Write and solve an equation to find x. Show your working.

a $6x - 2$ $5x + 4$

$x =$ _____

b
$4x + 3$
6
$7x - 6$
Units are in cm. $x =$ _____

What is the area of this rectangle? _____ cm^2

(C) Find the number

Write an equation and use it to answer the question.

a Three times a number minus 8 gives the same answer as subtracting the number from 20. What is the number?

Equation _____   Number _____

(D) True or false?

Six students started to solve the equation $4x + 7 = 2x + 11$ in different ways.
For each student's statement, write true or false.

a Jack wrote
$4x + 7 = 2x + 11$
so $4x + 2x = 11 + 7$ _____

b Ben wrote
$4x + 7 = 2x + 11$
so $11x = 13x$ _____

c Alice wrote
$4x + 7 = 2x + 11$
so $2x + 7 = 11$ _____

d Nick wrote
$4x + 7 = 2x + 11$
so $4x = 2x + 4$ _____

e Hayley wrote
$4x + 7 = 2x + 11$
so $2x = 4$ _____

f Jessie wrote
$4x + 7 = 2x + 11$
so $^-4 = ^-2x$ _____

     How did you find this?   EASY   OK   HARD

33 Solving non-linear equations

Let's look at ...
- solving non-linear equations mentally, using a calculator or using trial and improvement

These are the points you need to know.

✓ **Non-linear equations** have terms with indices greater than 1, such as x^2, x^3, ...
We can solve some non-linear equations to get an exact answer.

Example $x^2 - 10 = 39$
$$x^2 = 39 + 10$$
$$x^2 = 49$$
$$x = \sqrt[\pm]{49}$$
$$= +7 \text{ or } ^-7$$

 When you find the square root, give the positive and negative solutions.

✓ Harder non-linear equations are solved using **trial and improvement** or using a calculator, spreadsheet or graph plotting software. The answer is not usually exact.

Example $m^2 + 5 = 27$
$$m^2 = 27 - 5$$
$$= 22$$
$$m = \sqrt[\pm]{22}$$
$$= +4·69 \text{ (2 d.p.) or } ^-4·69 \text{ (2 d.p.)}$$

(A) Hamish's questions

Solve each of these equations **mentally**. Show your working.

a
$$y^2 + 21 = 102$$

b
$$a^2 - 52 = 92$$

c
$$8 = \frac{72}{c^2}$$

(B) Victor's choice

Use **your calculator** to find the solutions to these. Round to 2 d.p. if necessary.

a
$$m^2 + 6 = 44·44$$

b
$$x^3 - 97 = 2100$$

c
$$2·3z^3 = 1·1776$$

(C) Natalie's method

a Natalie wants to solve $x^2 - x = 60$ using **trial and improvement**.
She has started this table.

We can see from Natalie's table that x must be between
8·2 and 8·3.

i Try each of these values for x.
Fill in the table for each.
- 8·25 - 8·27 - 8·26

ii Use your answers from **a** to solve $x^2 - x = 60$ to
2 d.p. $x = \_\_\_\_$

Try	$x^2 - x$	Comment
$x = 8$	56	too small
$x = 9$	72	too big
$x = 8·4$	62·16	too big
$x = 8·3$	60·59	too big but close
$x = 8·2$	59·04	too small but close

b Use the trial and improvement method to solve
$y^3 + y = 40$ to 2 d.p.
$y = \_\_\_\_$

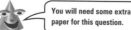 You will need some extra paper for this question.

How did you find this? EASY OK HARD

34 Algebra and proportion

Let's look at ...
● variables which are in direct proportion

✓ When two variables are directly proportional, the **ratio** of corresponding values is always the same.

p	1	2	3	4	5
q	2	4	6	8	10

$$\frac{q}{p} = \frac{2}{1} = \frac{4}{2} = \frac{6}{3} = \frac{8}{4} = \frac{10}{5}$$
$$q = 2p$$

This information can be plotted on a **graph**.
If the variables are directly proportional the points will lie in a straight line through the origin.

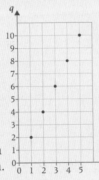

✓ We can use **algebra** to solve problems if the variables are directly proportional.

Example If we make coffee by mixing 2 spoons of sugar (s) for every 5 spoons of coffee (c), then the ratio $\frac{s}{c}$ is constant.

To work out how much sugar is needed for 45 spoons of coffee we can use algebra.

$$\frac{s}{45} = \frac{2}{5}$$
$$s = \frac{2}{5} \times 45$$
$$= \textbf{19 spoons}$$

These are the points you need to know.

(A) Stuck firm

A specialist glue is made by mixing 2 parts of glue one with 3 parts of glue two.

a Complete this table.

Glue 1 (ml)	2	4	6	8	10
Glue 2 (ml)					

b Is the amount of glue 1 directly proportional to the amount of glue 2? _____ Justify your answer using ratio _____

_____

c Draw a graph of glue 1 versus glue 2.

d Is your graph a straight line? _____ Explain why or why not.

_____

e Write the relationship between glue 1 (f) and glue 2 (s) as an equation. $s = $ _____

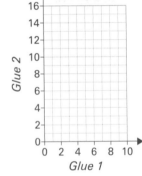

Mixing glue

Glue 2 / Glue 1

(B) Chasing shadows

Jay's group did an experiment to test if the height of objects is directly proportional to the length of their shadows, at a certain time. They made these measurements at midday on the same day.

Object height (cm)	50	110	220	190	280	80	30	160	250
Shadow length (cm)	40	90	176	160	220	54	25	122	116

a Draw a graph of this data on the grid.

b Give a possible reason why the points don't lie in an **exact** straight line. _____

c Do you think that object height and shadow length are directly proportional at a certain time? _____
Explain _____

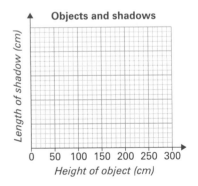

Objects and shadows

Length of shadow (cm) / Height of object (cm)

(C) Mix it up

Use algebra to find the answer.

Purple paint is made by mixing 7 parts of red with 4 parts of blue. How much blue paint would be mixed with 280 ml of red?

How did you find this?   EASY   OK   HARD

35 Collecting like terms

Let's look at ...
- simplifying expressions by collecting like terms

✔ We can **simplify expressions** by **collecting like terms**.

Examples
- $3n + 4m - n - 3m = 3n - n + 4m - 3m$
  $= 2n + m$

  like terms — like terms

  Write like terms next to each other

  The sign before the term moves with it

- $8q + 2(p - 3q) = 8q + 2p - 6q$
  $= 2q + 2p$

  Expand the brackets first

- $6x^2 - 2x + 3x^2 + 3x = 6x^2 + 3x^2 - 2x + 3x$
  $= 9x^2 + x$

  x and x^2 are **not** like terms so cannot be added or subtracted.

These are the points you need to know.

A Match it up or circle it

a Draw a line to match each expression on the left with its simplest form on the right.

$3x^2 + x + 2x^2 + 4x$ •   • $8x^2 - 5x$
$x^2 - 3x - 2x + 7x^2$ •   • $5x^2 - 5$
$7x^2 + 1 - 2x^2 - 6$ •   • $5x - 1$
$2(x - 2) + 3(x + 1)$ •   • $5x^2 + 5x$
$5(x - 2) + 3(x + 1)$ •   • $8x - 15$
$7(x - 2) - (1 - x)$ •   • $8x - 7$

b Circle the odd one out in each stack.

i

$e^2 - 2 + e^2 - 3$
$7e^2 - 4 - 5e^2 - 1$
$3e^2 - 5 - 2e^2$
$10e^2 + 10 - 8e^2 - 15$

ii

$5c + 4 - 2c$
$2(c + 2) + c$
$3c^2 + 4 + 3c - 3c^2$
$6c + 4 - 2c - 1$

B Round the outside

For **a** and **b** write an expression in its simplest form for the perimeter (P).

For **c** write an expression for the missing lengths.

a
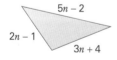

$5n - 2$
$2n - 1$
$3n + 4$

$P = $ _____
 $= $ _____

b

$5m + 1$
$3m - 4$

$P = $ _____
 $= $ _____

c
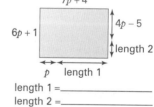

$7p + 4$
$6p + 1$
$4p - 5$
length 2
p length 1

length 1 = _____
length 2 = _____

C Pyramids

In these pyramids each expression is found by adding the two expressions below.
Complete each pyramid.

There are many different ways to complete **c**.

a
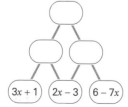

$3x + 1$ $2x - 3$ $6 - 7x$

b
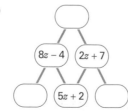

$8z - 4$ $2z + 7$
$5z + 2$

*c
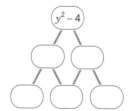

$y^2 - 4$

36 Multiplying and dividing expressions

Let's look at ...
- simplifying expressions by cancelling and using the index laws

These are the points you need to know.

✓ **Remember**
- $4 \times 2c = 8c$
- $3m \times 6m = 3 \times m \times 6 \times m$
$$= 3 \times 6 \times m \times m$$
$$= 18 \times m^2$$
$$= 18m^2$$

✓ We can simplify expressions by **cancelling**.   *Examples* $\frac{8m^1}{m^1} = 8$ $\frac{18x^2}{12x} = \frac{{}^3\cancel{18} \times x \times x \times x^1}{{}_2\cancel{12} \times x_1}$
$$= \frac{3x}{2}$$

✓ We can also simplify expressions using **the index laws**.
- When multiplying numbers with indices, we **add** the indices.   *Example* $2y^2 \times y^3 = 2 \times y^2 \times y^3$
$$= 2 \times y^{2+3}$$
$$= 2y^5$$

- When **dividing** numbers with indices, we **subtract** the indices.   *Example* $\frac{3k^5}{k^3} = 3k^{5-3}$
$$= 3k^2$$

(A) What do you think?

Simplify these equations, then shade the answer in the grid. The shaded answers will form a number. This number is the percentage of your brain which is water.

15m	m^2	$6m^2$	$8mn$	$12mn$	$30m^{10}$	m
$6m^8$	$30m^{35}$	^-8m	$11m^{10}$	$32m^2$	$\frac{28m}{7}$	m^8
$10m$	$7m^3$	m^4	$5m^2$	$24m$	$\frac{4m^2}{3}$	$4m$
$12m^2$	$2m$	$4m^2$	^-m	m^6	m^{-2}	$10m^3$
$10m^2$	$8m$	$6m^6$	m^3	^-12m	m^7	$\frac{7m}{5}$

a $3 \times 5m$ _____

b $4m \times 6$ _____

c $^-2 \times 4m$ _____

d $3m \times {}^-4$ _____

e $2m \times 3m$ _____

f $4m \times 8m$ _____

g $6m \times 2n$ _____

h $\frac{m^2}{m}$ _____

i $\frac{35m}{25}$ _____

j $m \times m^3$ _____

k $\frac{m^4}{m^2}$ _____

l $m^{10} \div m^3$ _____

m $\frac{28m^2}{7m}$ _____

n $\frac{12m^5}{3m^3}$ _____

o $6m^2 \times m^4$ _____

*p $2m \times 5m^2$ _____

*q $6m^3 \times 5m^7$ _____

*r $\frac{16m^5}{12m^3}$ _____

Your brain is _____ % water.

(B) Odd one out

Circle the odd one out in each stack.

a
$3 \times 5n$
$5 \times 3n$
$^-3 \times {}^-5n$
$15n \times n$

b
$3b \times 4b$
$2 \times 6b^2$
$4b \times 3b^2$
$6b \times 2b$

*c
$\frac{6p^4}{4p}$
$3p^5 \div 2p^2$
$\frac{15p^6}{10p^2}$
$\frac{12p^{10}}{8p^7}$

(C) Lots of fish

a Write an expression for the area (A) of each of these fish ponds.

i
5x, 3x
$A =$ _____

ii
3, 6x
$A =$ _____

iii
6x, 4x, 10x
$A =$ _____

*b Write an expression for the length labelled l in each of these.

i
4x
l
$l =$ _____
area = 24x

ii
$l =$ _____
l
8x
area = $12x^2$

How did you find this?   EASY   OK   HARD

37 Writing expressions

Let's look at ...
● writing expressions and using them to solve problems

These are the points you need to know.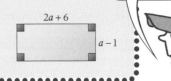

✓ We can **write expressions**.

Example An expression for the perimeter of this rectangle is
$$2(2a + 6 + a - 1) = 2(3a + 5)$$
$$= 6a + 10$$

2a + 6

a − 1

A Sweet tooth

Yasmin has a large bag of sweets.

In the bag are $8x - 2$ sweets.

bag of sweets

8x − 2

piles of loose sweets

a Yasmin puts the sweets into two piles.
There are $3x - 1$ sweets in the first pile.
How many sweets are in the second pile? ____

3x − 1   ?

b Yasmin puts all the sweets back in her bag.
Then she takes them out and puts them in **two equal piles**.
How many sweets are in each pile? ____

?   ?

c Yasmin puts all the sweets back in her bag again, then she puts them in these two different piles.
There are **12** sweets in the first pile.
How many sweets are in the second pile? ____

x + 3   7x − 5

B Number puzzles

You can often use algebra to show why a number puzzle works.

Fill in the missing expressions.

a Example: Algebra

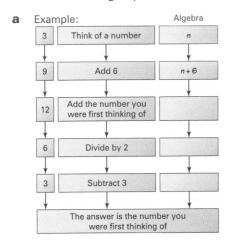

3	Think of a number	n
9	Add 6	n + 6
12	Add the number you were first thinking of	
6	Divide by 2	
3	Subtract 3	
	The answer is the number you were first thinking of	

b Example: Algebra

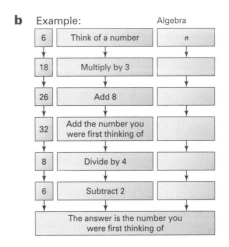

6	Think of a number	n
18	Multiply by 3	
26	Add 8	
32	Add the number you were first thinking of	
8	Divide by 4	
6	Subtract 2	
	The answer is the number you were first thinking of	

C Perimeter and area

a Write simplified expressions for the area (A) and perimeter (P) of this rectangle. $A =$ _____ $P =$ _____

7a

3b

*b A different rectangle has area $18a^2$ and perimeter $22a$.
What are the dimensions of this rectangle?   _____ by _____

*c i Write a simplified expression for the perimeters of these rectangles.
perimeter of A = _____   perimeter of B = _____

*ii If the perimeters of A and B are the same, find the value of x.

3x   2x + 1
x   A   B   x + 1

$x =$ _____

How did you find this?   EASY   OK   HARD

38 Factorising

Let's look at ...
● factorising expressions

✓ **Factorising** an expression is the inverse of multiplying out a bracket.

✓ When we **factorise** an expression we put the highest common factor of the terms outside the bracket.

Examples $12x - 16$ is factorised as $4(3x - 4)$ **4 is the HCF of 12x and 16**
        $8y + 24w$ is factorised as $8(y + 3w)$ **8 is the HCF of 8y and 24w**
        $m^3 + m^2 + 3m$ is factorised as $m(m^2 + m + 3)$ **m is the HCF of m^3, m^2 and 3m**

These are the points you need to know.

✓ We can factorise using a grid to help.

Example $12x - 8$ can be factorised like this.

| 12x | ⁻8 |

| 4 | 12x | ⁻8 |

4 is a common factor
of 12x and ⁻8.

$3x$
| 4 | 12x | |

$4 \times 3x = 12x$
or $\frac{12x}{4} = 3x$

$3x$ ⁻2
| 4 | 12x | ⁻8 |

$4 \times ⁻2 = ⁻8$
or $\frac{⁻8}{4} = ⁻2$

So $12x - 8 = 4(3x - 2)$

✓ Always check your factorising by multiplying out the brackets.

(A) Find the pairs

$3x + 27 = 3(x + 9)$ so $3x + 27$ and $3(x + 9)$ are a pair.
This pair has been circled in the grid. Circle all of the
other pairs.

3x + 27	3(x + 9)	5(x + 2)	2(x − 6)	2x − 8
3(x + 4)	3x + 12	5x + 10	2x − 12	2(x − 4)
3x + 9	5x + 25	5(x + 5)	2(x − 2)	2x − 4
3(x + 3)	3x + 6	3(x + 2)	2x − 2	2(x − 1)

(B) Have some help

Complete these.

a $4x + 8 = 4(\underline{\quad} + \underline{\quad})$
b $3n - 12 = 3(\underline{\quad} - \underline{\quad})$
c $5y + 5 = 5(\underline{\quad} + \underline{\quad})$
d $25a - 15 = 5(\underline{\quad} - \underline{\quad})$
e $16x + 12 = \underline{\quad}(4x + 3)$
f $24 + 8n = \underline{\quad}(3 + n)$
g $4 + 10x = \underline{\quad}(2 + \underline{\quad})$
h $18n - 12 = \underline{\quad}(\underline{\quad} - 2)$
i $3x + 3y = 3(\underline{\quad} + \underline{\quad})$
j $6a + 12b = \underline{\quad}(a + \underline{\quad})$
k $3y^2 + y = y(\underline{\quad} + \underline{\quad})$
l $x^2 - 5x = x(\underline{\quad} - \underline{\quad})$

(C) On your own

Factorise these.

a $2x + 6 = \underline{\quad}$
b $12y - 6 = \underline{\quad}$
c $15a + 10 = \underline{\quad}$
d $14 + 21n = \underline{\quad}$
e $35a - 25 = \underline{\quad}$
f $21x + 18 = \underline{\quad}$
g $36 - 30a = \underline{\quad}$
h $x^2 + 3x = \underline{\quad}$
*i $a^3 + a^2 = \underline{\quad}$
*j $5y^2 + 5y = \underline{\quad}$
*k $6n^2 + 3n = \underline{\quad}$
*l $4a^3 - a^2 = \underline{\quad}$

(D) The best and the worst

Each of the expressions in this table has
been factorised by Annie, Bob and Carla.

For each expression

i   one student has the answer **wrong** –
    cross this out

ii   one student has the **best** answer – circle this.

The first one is done for you.

This answer is wrong.

Expression	8x + 4	6a − 12	48x + 24	45 − 15n	* $y^3 + y^2$	* $6n^3 − n^2$
Annie	8(x + 1)	3(a − 4)	12(x + 2)	5(9 − 3n)	$y^3(1 + y)$	$n^2(6n − 1)$
Bob	2(4x + 2)	6(a − 2)	6(8x + 4)	15(3 − n)	$y(y^2 + y)$	$6n^2(n − 1)$
Carla	4(2x + 1)	3(2a − 4)	24(2x + 1)	5(9 − 5n)	$y^2(y + 1)$	$n(6n^2 − n)$

This is the best answer because 4 is the HCF of 8 and 4.

How did you find this?   **EASY**   **OK**   **HARD**

39 Adding and subtracting algebraic fractions

Let's look at ...
● adding and subtracting algebraic fractions by finding a common denominator

These are the points you need to know.

✓ We can **add and subtract algebraic fractions** in the same way that we add and subtract fractions in arithmetic.

Example   **Arithmetic**
$$\frac{3}{4} + \frac{2}{5} = \frac{3 \times 5}{20} + \frac{2 \times 4}{20}$$

find a common denominator and make equivalent fractions.

Algebra
$$\frac{4}{m} + \frac{p}{n} = \frac{4 \times n}{m \times n} + \frac{p \times m}{n \times m}$$

$$= \frac{15 + 8}{20}$$

$$= \frac{4n + pm}{mn}$$

$$= \frac{23}{20}$$

$$= 1\frac{3}{20}$$

A Starting slowly

Complete these.

a $\frac{2}{9} + \frac{5}{9} = \boxed{}$

b $\frac{4}{x} + \frac{3}{x} = \boxed{}$

c $\frac{6}{y} - \frac{2}{y} = \boxed{}$

d $\frac{y}{5} + \frac{2y}{5} = \boxed{}$

e $\frac{1}{2} + \frac{1}{3} = \frac{\boxed{}}{6} + \frac{\boxed{}}{6} = \boxed{}$

f $\frac{2}{3} - \frac{5}{8} = \frac{\boxed{}}{\boxed{}} - \frac{\boxed{}}{\boxed{}} = \boxed{}$

g $\frac{3}{a} + \frac{4}{b} = \frac{\boxed{}}{ab} + \frac{\boxed{}}{ab} = \frac{\boxed{}}{ab}$

h $\frac{c}{3} + \frac{d}{4} = \frac{\boxed{}}{\boxed{}} + \frac{\boxed{}}{\boxed{}} = \frac{\boxed{}}{\boxed{}}$

*i $\frac{p}{q} + \frac{r}{s} = \frac{\boxed{}}{\boxed{}} + \frac{\boxed{}}{\boxed{}} = \frac{\boxed{}}{\boxed{}}$

B Match it up

Draw a line to match each calculation in the top row with its answer in the bottom row.

$\frac{x}{2} + \frac{y}{5}$	$\frac{2}{x} + \frac{5}{y}$	$\frac{x}{2} - \frac{y}{5}$	$\frac{2}{x} - \frac{y}{5}$	$\frac{2}{x} + \frac{y}{5}$

$\frac{5x - 2y}{10}$	$\frac{5x + 2y}{10}$	$\frac{2y + 5x}{xy}$	$\frac{10 + xy}{5x}$	$\frac{10 - xy}{5x}$

C On your own

Add and subtract these.

a $\frac{a}{3} + \frac{a}{7} =$ _____

b $\frac{4}{b} + \frac{3}{c} =$ _____

*c $\frac{w}{x} + \frac{y}{z} =$ _____

D True or false?

Decide whether each of these is true (T) or false (F).

a $\frac{6}{p} + \frac{7}{p} = \frac{42}{p}$ _____

b $\frac{q}{2} + \frac{r}{7} = \frac{7q + 2r}{14}$ _____

c $\frac{s}{3} - \frac{t}{10} = \frac{3t - 10s}{30}$ _____

d $\frac{4}{u} + \frac{5}{v} = \frac{4v + 5u}{uv}$ _____

e $\frac{2m}{3} + \frac{m}{4} = \frac{11m}{12}$ _____

f $\frac{2a}{b} - \frac{3c}{d} = \frac{2ab - 3cd}{bd}$ _____

*E Card games

Majorie has these five cards. $\boxed{\frac{2}{3}}$ $\boxed{\frac{a}{b}}$ $\boxed{\frac{2}{a}}$ $\boxed{\frac{3}{b}}$ $\boxed{\frac{a}{2}}$

a She adds two cards and gets $\frac{a+3}{b}$. Which cards did she add? $\boxed{} + \boxed{}$

b She subtracts one card from another and gets $\frac{2b - 3a}{ab}$. Which cards? $\boxed{} - \boxed{}$

c She adds three cards and gets $\frac{4a + 12 + 3a^2}{6a}$.
One card is $\boxed{\frac{2}{a}}$. What are the other two cards? $\boxed{\frac{2}{a}} + \boxed{} + \boxed{}$

40 Substituting into expressions

Let's look at ...
● substituting values for the unknown into expressions

✓ We **find the value of an expression** by substituting values for the unknown. We follow the rules for **order of operations**.

Examples When $p = 5$

$$\frac{3p^2}{10} = \frac{3 \times 5^2}{10}$$
$$= \frac{3 \times 25^5}{10^2}$$
$$= \frac{15}{2}$$
$$= 7\frac{1}{2}$$

$$4(p-2)^2 = 4(5-2)^2$$
$$= 4(3)^2$$
$$= 4 \times 9$$
$$= 36$$

These are the points you need to know.

(A) Quick questions

a When $p = 5$, find the values of these.

i $3p$ ____   **ii** $2p - 3$ ____   **iii** $4(p + 6)$ ____   **iv** $3(p - 2)^2$ ____   **v** $\frac{4p^2}{5}$ ____

b When $q = {}^-2$, find the values of these.

i $4q$ ____   **ii** $3q - 1$ ____   **iii** $2q + 7$ ____   **iv** $q^2 + 6$ ____   **v** $3(q - 3)^2$ ____

(B) Match it up

a Join pairs of algebraic expressions that have the **same value** when $a = 4$, $b = 3$ and $c = 5$. One pair is joined for you.

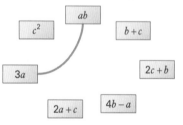

b Join pairs of expressions that have the same value when $x = y = z$.

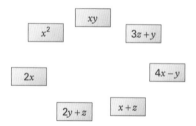

(C) Sheep pens

Farmer McGregor builds a series of holding pens from sections of fence.

The number of fence sections required for pen p is $p^2 + 3p$. How many fence sections are required for

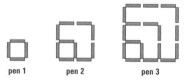

pen 1   pen 2   pen 3

a pen 4? _____   **b** pen 10? _____   **c** pen 30? _____

*(D) Odd or even?

y is an odd number.
Copy each of these into one of the boxes.

a $3y$ **b** y^2 **c** $2y$

d $5y + 1$ **e** $\frac{y-1}{2}$ **f** $(y + 2)(y - 2)$

Must be odd	Must be even	Can't tell

How did you find this?   EASY   OK   HARD

41 Substituting into formulae

Let's look at ...
- substituting into a formula to find unknown values

 When we substitute values for unknowns into a **formula** we sometimes need to solve an equation.

These are the points you need to know.

> *Example* The formula for finding speed, S, in m/sec, is $S = \frac{D}{T}$ where D is distance in metres and T is time in seconds.
> If $S = 22$ m/s and $T = 30$ sec, then
> $$22 = \frac{D}{30}$$
> $$22 \times 30 = D$$
> $$D = 660 \text{ m}$$

A Place your order

This is the formula for the price, £p, of f pieces of fish and c scoops of chips at a chip shop.

$$p = 2f + 1\cdot5c$$

Calculate the price paid by each of these families.

a The Robertsons had 4 fish and 2 scoops of chips. $p = $ ____

b The O'Learys ordered 7 fish and 4 scoops of chips. $p = $ ____

c The McFedries ate 11 fish and 5 scoops of chips. $p = $ ____

B Parallelograms

The formula for the area of a parallelogram is $A = bh$.

Calculate these. Show your working.

a
Find A if $b = 20$ cm, $h = 17$ cm.

b
Find A if $b = 1\cdot5$ m, $h = 0\cdot4$ m.

c
Find h if $A = 0\cdot42$ m², $b = 0\cdot7$ m.

C Eggs–actly

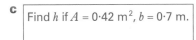
Give your answers to 2 d.p.

The mass in grams, w, of an egg can be calculated using this formula.

$$w = \frac{2l^3}{7}$$ where l is the length in cm of the egg

a Find the masses of these eggs. Round your answers to 2 d.p.

   i $l = 4$ cm, $w = $ _____   **ii** $l = 5\cdot2$ cm, $w = $ _____   **iii** $l = 7\cdot8$ cm, $w = $ _____

*b An egg weighs 40 g. How long is it? _____ Show your working.

* D Pythagoras

Pythagoras' formula for working out the length of side c in a right-angled triangle is:

$$c = \sqrt{a^2 + b^2}$$

Find c for these triangles. Round your answers to 2 d.p. when necessary.

a $a = 3$ cm, $b = 4$ cm _____   **b** $a = 6$ cm, $b = 7$ cm _____

c $a = 1\cdot7$ cm, $b = 5\cdot4$ cm _____   **d** $a = 0\cdot3$ m, $b = 0\cdot14$ m _____

How did you find this?   EASY   OK   HARD

43

42 Changing the subject of a formula

Let's look at ...
● rearranging formulae to change the subject

✓ Sometimes we need to **change the subject of a formula**.

Example $C = 2\pi r$ where C is the circumference of a circle
r is the radius
To make r the subject, use inverse operations.

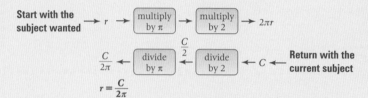

These are the points you need to know.

(A) How far?

s is the speed in km/h, d is distance in km and t is time in hours.

a Make d the subject of $s = \frac{d}{t}$.
Use the function machines to help.

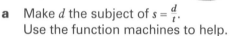

so $d =$ _____

b Use your new formula to find d if
i $s = 90$ km/h and $t = 4$ h $d =$ _____
ii $s = 53$ km/h and $t = 1·5$ h $d =$ _____

(B) Fill the cone

V is the volume of a cone, A is the area of its base and h is its height.

a Make h the subject of $V = \frac{Ah}{3}$.

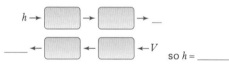

so $h =$ _____

b Find h if
i $A = 60$ mm^2 and $V = 160$ mm^3 $h =$ _____
ii $A = 41$ cm^2 and $V = 164$ cm^3 $h =$ _____

(C) Try these

Rearrange these formulae to give the subject shown.

a $C = \pi D$ $D =$ _____

b $d = l + 3t$ $l =$ _____ $t =$ _____

c $I = \frac{PRT}{100}$ $P =$ _____ $R =$ _____ $T =$ _____

(D) Match it up

Each formula in the top row has been rearranged to form a new formula in the bottom row. Draw a line to match equivalent formulae.

$d = 2ef$ $d = \frac{e+f}{2}$ $d = 2(e+f)$ $d = \frac{ef}{2}$

$e = 2d - f$ $e = \frac{2d}{f}$ $e = \frac{d}{2f}$ $e = \frac{d}{2} - f$

***E* Rearrange it**

a The subject of the equation below is t.

$= 2(r + s)$

Rearrange the equation to make r the subject.   _____

b Rearrange the equation $e = f^2 - g$ to make f the subject. _____

How did you find this?   EASY   OK   HARD

43 Finding formulae

Let's look at ...
● using given information to write formulae

These are the points you need to know.

 We can **write formulae** using given information.

Example When cubes are joined in a line we can count the number of square faces.

2 cubes
10 faces

3 cubes
14 faces

4 cubes
18 faces

To find a formula for the number of square faces on a line of n cubes we start by drawing a table.

The number of faces forms the sequence 10, 14, 18, 22, 26, ...

Each number is **4** more than the one before.

The formula will be

$$f = 4 \times n + ?$$

We can find what **?** is by looking at the sequence.

1st term $10 = 4 \times 2 + 2$
2nd term $14 = 4 \times 3 + 2$ and so on

So the formula is $f = 4n + 2$.

Number of cubes (n)	2	3	4	5	6	...
Number of faces (f)	10	14	18	22	26	...

 A *Algebra and area*

Show, using algebra, that the area of the dark shaded part of this diagram is $A = 5ab$.

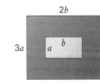

2b

3a a b

 B *How many edges?*

Carolyn was studying 3-D shapes which have at least one vertex (corner).
She decided to count the number of faces, vertices and edges of five shapes.

a Complete Carolyn's table for her.

Shape		Number of faces (f)	Number of vertices (v)	$f+v$	Number of edges (e)
Cube		6	8	14	12
Triangle-based pyramid					
Cone					
Triangular prism					
Hexagonal prism					

b Find a formula for the number of edges (e) in each of these 3-D shapes.

$e = $ _____

c Does your formula work for

i a square-based pyramid? _____   **ii** a cylinder? _____

---

44 Generating sequences

Let's look at ...
- using term-to-term rules and position-to-term rules to generate sequences

✔ Sequences can be generated from **term-to-term** rules.
    Example   **First term** 3   **Rule** add 5   generates 3, 8, 13, 18, 23, ...

✔ Sequences can also be generated from expressions for the *n*th term.
We call this a **position-to-term** rule.
This is sometimes written as $T(n)$.

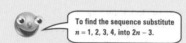 $T(n)$ is the *n*th term.

    Example   The rule for the *n*th term of a sequence is $T(n) = 2n - 3$.
             This generates the sequence

$$\begin{array}{cccc} {}^{-}1, & 1, & 3, & 5... \\ \uparrow & \uparrow & \uparrow & \uparrow \\ n=1 & n=2 & n=3 & n=4 \end{array}$$

To find the sequence substitute $n = 1, 2, 3, 4$, into $2n - 3$.

These are the points you need to know.

             The difference between terms is 2, which is the number multiplying *n*.

✔ **Linear sequences** always have a constant difference between terms.

(A) What comes next?

a   Write down the first six terms of these sequences

	first term(s)	term-to-term rule	
i	3	multiply by 2	_____
ii	7	subtract 3	_____
iii	2	add 0·4	_____
iv	5	add consecutive numbers 1, 2, 3, 4, ...	_____
v	2, 3	add the previous two terms	_____

b   Which of these sequences are linear? _____ Explain how you can tell. _____

_____

(B) Table of terms

Complete this table.

Rule	$T(1)$	$T(2)$	$T(3)$	$T(4)$	$T(5)$	...	$T(20)$
$T(n) = 30 - 2n$	28					...	
$T(n) = 3n + 0·5$						...	
$T(n) = \frac{n}{10}$						...	

*(C) Twelve terms

$$T(3) = T(2) - T(1), \qquad T(4) = T(3) - T(2), ...$$

a   Write down the first twelve terms of this sequence if $T(1) = 1$ and $T(2) = 3$.

_____

b   Look closely at these twelve terms. What pattern can be seen? _____

_____

           How did you find this?   **EASY**   **OK**   **HARD**

45 Describing and continuing sequences

Let's look at ...
- **continuing a variety of sequences**
- **describing linear sequences**

These are the points you need to know.

✓ We can often **predict how a sequence continues** if we are given the first few terms.

Example 5, ⁻10, 20, ⁻40, 80, ...
Each term in this sequence is the previous term multiplied by ⁻2 to give the next term.
So we predict the next term is 80 × ⁻2 = ⁻160.

✓ We can **describe linear sequences** by looking at the rule for the nth term.

Example $T(n) = 3n + 1$ gives the sequence 4, 7, 10, 13, ...
Each term is one more than a multiple of 3.
It is ascending with a common difference of 3.

A Continue it

First term 4, add 2 is a term-to-term rule.

a Predict the next three terms of these sequences and write down the term-to-term rule.

i 0·1, 0·5, 0·9, 1·3, _____, _____, _____ . Rule: _____

ii 4, ⁻40, 400, ⁻4000, _____, _____, _____ . Rule: _____

iii 192, 96, 48, 24, 12, _____, _____, _____ . Rule: _____

iv 2, 3, 5, 8, 13, 21, _____, _____, _____ . Rule: _____

b Continue this sequence in **two** different ways. Explain the rule for each.

2, 4, 6, _____, _____, _____ . Rule: _____

2, 4, 6, _____, _____, _____ . Rule: _____

B Match it

Draw a line to match each sequence with its description.

$T(n) = 5n$ • • All terms 2 less than a multiple of 7. Starts at 5 and ascends.
$T(n) = 5n + 2$ • • Starts at 7. Ascending terms with common difference of 5.
$T(n) = 7n - 2$ • • Starts at 35. Descending terms with common difference of 7.
$T(n) = 40 - 5n$ • • Ascending multiples of 5.
$T(n) = 42 - 7n$ • • Multiples of 5. Starts at 35 and descends.

C Describe it

Describe the terms of the sequence given by each of these.
The first one is done for you.

a $T(n) = 4n - 1$ *Each term is one less than a multiple of 4. It starts at 3 and ascends*.

b $T(n) = 6n$ _____

c $T(n) = 6n + 1$ _____

d $T(n) = 66 - 6n$ _____

D Puzzle it

Anna made up a sequence.

a She said 'The numbers in my sequence are all multiples of 4.'
What might her first term and rule be? Give two possible answers.

First term _____ . Rule: _____

or First term _____ . Rule: _____

46 Quadratic sequences

Let's look at ...
- generating quadratic sequences
- recognising quadratic sequences by finding their second differences

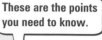

These are the points you need to know.

✓ **Quadratic sequences** have rules with a squared term as the highest power.

Example $T(n) = n^2 + 1$

Position	1	2	3	4	5	...
Term $T(n)$	$1^2 + 1 = 2$	$2^2 + 1 = 5$	$3^2 + 1 = 10$	$4^2 + 1 = 17$	$5^2 + 1 = 26$	...

The quadratic sequence is 2, 5, 10, 17, 26.

✓ Quadratic sequences have a common **second** difference.

Example 2, 5, 10, 17, 26

3   5   7   9   ←——— 1st difference

2   2   2   ←——— 2nd difference is constant

(A) Table of terms

Complete this table.

Rule	$T(1)$	$T(2)$	$T(3)$	$T(4)$	$T(5)$	$T(6)$
$T(n) = n^2$	1	4				
$T(n) = n^2 + 2$	3					
$T(n) = 2n^2 + 4$						
$T(n) = n^2 + n$						
$T(n) = 10 - 2n^2$						
$T(n) = n^2 + 4n - 5$						

(B) Yay or nay?

a Fill in the missing numbers to find the second differences of these sequences.

i   6,   9,   14,   21,   30,   ...

   3   5   7   ☐

    2   2   ☐

ii   4,   6,   9,   14,   22,   ...

   2   3   ☐   ☐

    1   ☐   ☐

iii   3,   7,   13,   21,   31,   ...

   ☐   ☐   ☐   ☐

    ☐   ☐   ☐

b Which of the above sequences is quadratic? _____

How do you know? _____

(C) Spot the mistakes

In each of these tables one number in the bottom row is incorrect.
Circle the mistake.

a

Term number	1	2	5	10	n
Term	0	7	52	407	$2n^2 + n - 3$

*b

Term number	1	2	10	20	n
Term	⁻3	⁻98	⁻498	⁻1998	$2 - 5n^2$

(D) Take your time

Write down the first four terms of these sequences.

a $T(n) = \dfrac{n}{n^2 + 3}$ _____

b $T(n) = \dfrac{2n - 1}{n^2 + 1}$ _____

How did you find this?   **EASY**   **OK**   **HARD**

47 Sequences in practical situations

Let's look at ...
● continuing and writing rules for sequences in practical situations

These are the points you need to know.

✓ Sequences can be generated from practical situations.

Example

□ □ □

Shape 1   Shape 2   Shape 3

The expression for the number of circles in the *n*th shape is 4*n*.
Each time a new shape is drawn 4 new circles are added.
There are *n* lots of 4 where *n* is the shape number.

(A) Piles of tiles

Ruth makes a sequence of patterns with purple and white triangular tiles.

 pattern number 1

 pattern number 2

 pattern number 3

The rule for finding the number of tiles in pattern number *n* is: $\boxed{\text{number of tiles} = 1 + 3n}$

a   i   What does the 1 in the rule represent? _____
  ii   What does the **3n** represent? _____

b   Ruth makes **pattern number** 15 in her sequence. How many **white** tiles and how many **purple** tiles does she use?   _____ white and _____ purple tiles.

c   Ruth uses **151 tiles** altogether to make a pattern in her sequence.
    What is the number of the pattern she makes?   pattern number _____

(B) Snowflakes

Dominic made this pattern with some hexagonal shapes.
Each pattern has **1 white** hexagon in the middle.

a   Explain how the next pattern in the sequence is made.

    _____

b   Write the rule for finding the number of hexagons, *H*, in pattern number *n* in Dominic's sequence.   *H* = _____

pattern number 1   pattern number 2   pattern number 3

(C) Imagine

Jena uses some tiles to make a sequence of patterns.
The rule for finding the number of tiles, *T*, in pattern *N* in Jena's sequence is $T = 4N + 1$.
Draw what you think the first 3 patterns in Jena's sequence could be.

(D) Thomas's Tower

Alex is helping his little brother build a tower of blocks.

a   Complete this table.

Level	1	2	3	4	5
Number of blocks	7	12			

b   What is the term-to-term rule for the sequence the number of blocks makes?
    First term _____ Rule _____

 level 1

 level 2

 level 3

c   Write an expression for the number of blocks in level *n*.   Number of blocks = _____

d   How many blocks are needed for   **i** level 10? _____   **ii** level 20? _____

How did you find this?   **EASY**   **OK**   **HARD**

48 Finding the rule for the *n*th term

Let's look at ...
● writing a rule for the *n*th term of a linear sequence

These are the points you need to know.

We can find a **rule for the *n*th** term by finding the difference between consecutive terms.

Example

Term	13	19	25	31	37
Difference		6	6	6	6

The difference between consecutive terms is **6** so the *n*th term is of the form $T(n) = 6n + ?$

$T(1) = 13$ $6 \times 1 + 7 = 13$

$T(n) = 6n + 7$

Check by testing some more terms. $T(2) = 6 \times 2 + 7$ $T(3) = 6 \times 3 + 7$
$= 19 ✓$ $= 25 ✓$

(A) What's the rule?

Complete the difference table for each of these. Use this to find the *n*th term and the 50th term of each sequence.

a   6, 8, 10, 12, 14, ...

Term	6	8	10	12	14
Difference		2	2		

$T(n) = $ _____ $T(50) = $ _____

b   27, 32, 37, 42, 47, ...

Term	27	32			
Difference					

$T(n) = $ _____ $T(50) = $ _____

*c   4, 3, 2, 1, 0, ...

Term	4				
Difference					

$T(n) = $ _____ $T(50) = $ _____

*d   2·1, 2·2, 2·3, 2·4, 2·5, ...

Term					
Difference					

$T(n) = $ _____ $T(50) = $ _____

*(B) Which term is that?

Write a rule for each of the following sequences. Then find the required term.

a   6, 11, 16, 21, 26, ... $T(n) = $ _____
Which term is equal to 101? _____

b   8, 6, 4, 2, 0, ... $T(n) = $ _____
Which term is equal to ⁻90? _____

*(C) How many matches?

Sam made squares out of matches.

1 square   2 squares   3 squares

a   Complete this table.

Number of squares	1	2	3	4	5
Number of matches	4				

b   How many matches would Sam need to make 25 squares? _____

c   Matthew has 151 matches. How many squares can he make? _____

d   Ashleigh has 201 matches. Do any square patterns use exactly 201 matches? _____ How do you know? _____

How did you find this?   **EASY**   **OK**   **HARD**

49 Functions

Let's look at ...
● finding outputs and inputs of functions

✓ **Remember**: We can find the **output** of a function machine if we are given the input.

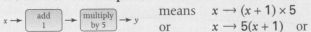

means $\quad x \rightarrow (x + 1) \times 5$

or $\qquad x \rightarrow 5(x + 1) \quad$ or $\quad y = 5(x + 1)$

These are the points you need to know.

✓ If the **output in a function machine** is 11, we work backwards to find the input. The input was 3.

✓ The **identity function** is $x \rightarrow x$. It maps every number onto itself.

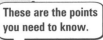

A What's missing?

Fill in the missing input and output numbers in the table.

a
$x \rightarrow$ [multiply by 3] $\rightarrow$ [subtract 2] $\rightarrow y$

Input	5	12		
Output			7	28

b
$x \rightarrow$ [divide by 4] $\rightarrow$ [add 7] $\rightarrow y$

Input	8	200		
Output			10	12

B Map it

Fill in the mapping diagrams for the functions given.

a $y = 2x - 1$ for $x = 1, 2, 3, 4, 5$

$2 \times \mathbf{1} - 1 = 1$ so 1 maps on to 1

b $x \rightarrow \frac{x+1}{2}$ for $x = 0, 1, 2, ^-1, ^-2$

c $x \rightarrow x$ for $x = ^-4, ^-1, 0, 2, 3 \cdot 5$

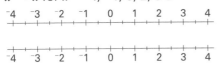

What special name does this function have? _____

What would $^-41 \cdot 9$ map on to? _____

C Shrink it

Each of these function machines can be replaced with a single-operation machine. Write down the single operation.

a $x \rightarrow$ [add 8] $\rightarrow$ [subtract 3] $\rightarrow y$ ☐

b $x \rightarrow$ [multiply by 6] $\rightarrow$ [divide by 2] $\rightarrow y$ ☐

D Match it

Find a function from the box that would give the same output as each of these.

a $1, 2, 3 \rightarrow$ [multiply by 5] $\rightarrow$ [add 5] $\rightarrow$ __

b $3, 5, 7 \rightarrow$ [add 2] $\rightarrow$ [multiply by 5] $\rightarrow$ __

c $2, 4, 6 \rightarrow$ [divide by 2] $\rightarrow$ [subtract 2] $\rightarrow$ __

A $y = 5x + 2$	**D** $y = \frac{x}{2} - 2$
B $y = \frac{x-2}{2}$	**E** $y = 5(x + 2)$
C $y = 5x + 5$	**F** $y = 5(x + 5)$

How did you find this? ⬡ EASY ⬡ OK ⬡ HARD

50 Inverse of a function

Let's look at ...
● using mapping diagrams and function machines to find the inverse of a function

✓ The **inverse of a function** can be found by doing the inverse operations in the reverse order.

Example

$x \rightarrow$ [add 4] $\rightarrow$ [multiply by 3] $\rightarrow 3(x+4)$ **function machine**

$\frac{x}{3} - 4 \leftarrow$ [subtract 4] $\overset{\frac{x}{3}}{\leftarrow}$ [divide by 3] $\leftarrow x$ **inverse function machine**

The inverse of $3(x+4)$ is $\frac{x}{3} - 4$

These are the points you need to know.

✓ On a mapping diagram the inverse function maps the output back to the input.

Example

$x \rightarrow 2x$
function

$x \rightarrow \frac{x}{2}$
inverse function

The inverse of multiplying by 2 is dividing by 2

The inverse of $x \rightarrow 2x$ is $x \rightarrow \frac{x}{2}$.

(A) Use a mapping diagram

a Fill in the mapping diagram to show this function and its inverse.

$x \rightarrow x - 2$ for $x = 0, 1, 2, 3, 4$

b What is the inverse function of $x \rightarrow x - 2$? _____

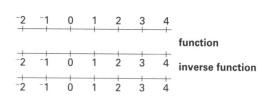

function

inverse function

(B) Use a function machine

Fill in the inverse function machine to find the inverse function.

a
$x \rightarrow$ [multiply by 3] $\rightarrow 3x$

$\leftarrow$ [ ] $\leftarrow x$

The inverse of $x \rightarrow 3x$ is _____ .

b
$x \rightarrow$ [divide by 4] $\rightarrow$ [subtract 3] $\rightarrow \frac{x}{4} - 3$

$\leftarrow$ [ ] $\leftarrow$ [ ] $\leftarrow x$

The inverse of $x \rightarrow \frac{x}{4} - 3$ is _____ .

c
$x \rightarrow$ [subtract 2] $\rightarrow$ [multiply by 7] $\rightarrow 7(x-2)$

$\leftarrow$ [ ] $\leftarrow$ [ ] $\leftarrow x$

The inverse of $x \rightarrow 7(x-2)$ is _____ .

d
$x \rightarrow$ [add 6] $\rightarrow$ [divide by 5] $\rightarrow \frac{x+6}{5}$

$\leftarrow$ [ ] $\leftarrow$ [ ] $\leftarrow x$

The inverse of $x \rightarrow \frac{x+6}{5}$ is _____ .

(C) Your choice

You may need some extra paper for your working.

Find the inverse function for these. You may choose the method you use.

a $x \rightarrow 4x$
Inverse _____

b $x \rightarrow x - 8$
Inverse _____

c $x \rightarrow 3x - 2$
Inverse _____

d $x \rightarrow 2(x + 1)$
Inverse _____

e $x \rightarrow \frac{x}{2} - 2$
Inverse _____

f $x \rightarrow 6 - x$
Inverse _____

How did you find this?   **EASY**   **OK**   **HARD**

51 Graphing functions

Let's look at ...
● plotting and interpreting graphs of the form $y = mx + c$
● rearranging straight-line equations to the form $y = mx + c$

✓ $y = mx + c$ is the **equation of a straight line**.
m is the **gradient** or steepness of the line.
c is the **y-intercept** (where the line cuts the y-axis).

If m is positive the line slopes ╱.
If m is negative the line slopes ╲.

✓ We can **rearrange equations into the form** $y = mx + c$.

Example $y - 2x + 3 = 0$
$\quad\quad\quad\quad y + 3 = 2x$ The inverse of subtracting $2x$ is adding $2x$.
$\quad\quad\quad\quad\quad y = 2x - 3$ The inverse of adding 3 is subtracting 3.

These are the points you need to know.

✓ Once the equation is in the form $y = mx + c$ we can plot the graph by constructing a table of values.

Example $y = 2x - 3$ **Choose 2 or 3 x-values** ⟶

x	0	1	3
y	⁻3	⁻1	3

(A) Two straight lines

a Complete these tables for the given equations.

i $y = \frac{1}{2}x + 3$

x	0	4	⁻4
y			

ii $y = \frac{1}{2}x - 1$

x	0	4	⁻4
y			

b Draw and label the lines $y = \frac{1}{2}x + 3$ and $y = \frac{1}{2}x - 1$.

c Will the point (10, 9) lie on the line $y = \frac{1}{2}x + 3$? _____
Explain. _____

d The point (⁻10, ?) lies on the line $y = \frac{1}{2}x - 1$. What is the missing y-coordinate? _____

e Make a comment about the similarities between the graphs of $y = \frac{1}{2}x + 3$ and $y = \frac{1}{2}x - 1$.
_____

(B) Match it up

These straight-line graphs all pass through the point (2, 2).

a Match each equation below with a line from the graph.

i $y = 2$ _____   **ii** $x = 2$ _____   **iii** $y = x$ _____
iv $y = \frac{1}{2}x + 1$ _____   **v** $y = ^{-}x + 4$ _____

b Does the line that has the equation $y = ^{-}2x + 6$ pass through the point (2, 2)? _____ Explain. _____

*c I want a line with equation $y = mx - 4$ to pass through the point (2, 2). What is the value of m? _____

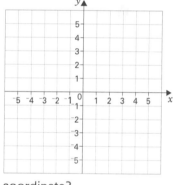

(C) Choose carefully

a Write each of these equations in the form $y = mx + c$.

A $y + 1 = 3x$ _____   **B** $2y = 8x + 4$ _____   **C** $\frac{1}{2}y = 2 + 3x$ _____   **D** $2y - 6x = 4$ _____

b Which of the equations in **a** have a gradient of 3? _____

c Three of these lines cut the y-axis at the same point. Circle them.

$y = 2x - 1$ $3x - y + 3 = 0$ $9y = 3x - 9$ $y + x + 3 = 0$ $\frac{y + 1}{3} = x$

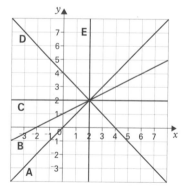

How did you find this?   **EASY**   **OK**   **HARD**

52 Gradient of a straight line

Let's look at ...
- finding the gradient, and the equation, of a straight line

✓ For a straight line, the change in y is **proportional** to the change in x.

$\frac{\text{change in } y}{\text{change in } x} = m$, the gradient of the line $m = \frac{y_2 - y_1}{x_2 - x_1}$ for any two points on the line.

> Always check that a positive slope has a positive value and a negative slope has a negative value for the gradient.

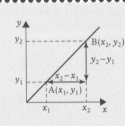

These are the points you need to know.

✓ To find the **equation of a line**, find the gradient, m, and the y-intercept, c. Then substitute these into $y = mx + c$ to give the equation.

Example The equation of this line is $y = 2x + 1$ because the gradient is 2 and it crosses the y-axis at 1.

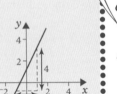

$\text{gradient} = \frac{\text{change in } y}{\text{change in } x} = \frac{4}{2} = 2$

(A) Find m

Find the gradient of l_1, l_2, l_3 and l_4.

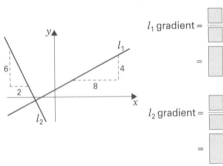

l_1 gradient = ☐/☐ = ☐

l_2 gradient = ☐/☐ = ☐

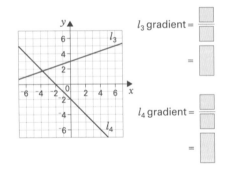

l_3 gradient = ☐/☐ = ☐

l_4 gradient = ☐/☐ = ☐

(B) Match it

a Name the lines that have a positive gradient. _____

b Name the lines that have a negative gradient. _____

c What is the gradient of line F? _____

d Match each equation with a line from the graph.

   i $y = 2x + 2$ _____   ii $y = {}^-x - 2$ _____   iii $y = {}^-2$ _____

   iv $y = \frac{1}{2}x + 2$ _____   v $y = x - 2$ _____   vi $y = -\frac{1}{2}x + 2$ _____

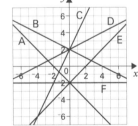

(C) Find the equations

Write down the gradients and the equations of lines R, S and T.

Line R Gradient _____ Equation _____

Line S Gradient _____ Equation _____

Line T Gradient _____ Equation _____

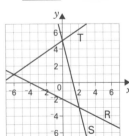

How did you find this? **EASY** **OK** **HARD**

53 Distance-time graphs

Let's look at ...
● interpreting and plotting distance–time graphs

✔ We can tell some things about how an object is moving by looking at its **distance–time graph**.

✔ The **gradient** of a distance–time graph tells us the speed. A steeper slope represents a faster speed.

Example Laura cycled at a steady pace.
She cycled 10 km in 30 mins. She stopped for 15 mins then cycled more slowly at a steady pace back to the start in 45 minutes.
She drew this graph.
We can use the graph to estimate how far she had cycled after 15 mins.
She had cycled about 5 km.

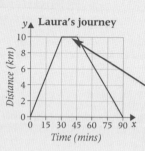

These are the points you need to know.

A horizontal line means Laura has stopped.

A Which walk is which?

 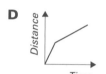

i Ginny walked quickly to the common, then slowly and steadily through the common. _____

ii Penny walked slowly to the post box, then ran home. _____

iii Carrie walked at a steady speed, stopped for a rest, then walked home again at the same speed. _____

*__iv__ Alex started slowly, then walked faster and faster until she was running. _____

B Cycling trip

Greg and Kieran set off on a weekend's camping trip. This graph shows their cycle journey to the campsite.

a What time did they arrive at the campsite? _____

b How far had they travelled before they first stopped for a rest? _____

c Kieran and Greg had to cycle slowly up a hill.
  i How long did this take? _____
  ii How long was the road up the hill? _____

d At the top of the hill they rested. How long for? _____

e During which section of their journey did they travel most quickly? From _____ km to _____ km.
Why do you think this was? _____

_____

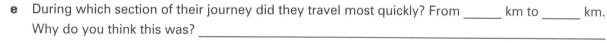

f Greg's father went in his car to drop the tent off for them. He left at 9:15 a.m. and travelled 18 km in 15 minutes. He stopped for 15 minutes to buy some new tent pegs. He reached the campsite 15 minutes later, travelling at a constant speed. He stayed half an hour to put up the tent then travelled back at a constant speed, arriving at 11 a.m.
Draw Greg's father's journey on the same grid.

How did you find this?   EASY   OK   HARD

54 Real-life graphs

Let's look at ...
● drawing, sketching and interpreting real-life graphs

✓ Sometimes we **sketch** a graph for a real-life situation, or interpret a sketch.

Example This graph shows the height of a ball above the ground when Todd threw it.

These are the points you need to know.

A Some apples please

Mrs King can buy her apples from the supermarket down the road or from the market across town.

a Mrs King walks to her local supermarket, then pays £2·50 per kg of apples. Draw the graph of the relationship between mass of apples and amount paid. The first point is plotted. Label this line 'supermarket'.

b To buy apples from the market Mrs King must first pay £6 for her return bus ticket, and then £1·50 per kg of apples. Graph this relationship on the same graph. Label the line.

c Where should Mrs King go if she needs

  i 4·5 kg of apples? _____   **ii** 9·1 kg of apples? _____

d For what mass of apples will the cost be the same? _____

B How deep?

a Water flowed steadily at the same rate into each of these containers. A depth (*d*) against time (*t*) graph was drawn for each.
Which graph represents each container? Explain your answer.

i

Graph _____ because
_____
_____
_____

ii

Graph _____ because
_____
_____
_____

iii

Graph _____ because
_____
_____
_____

*b Sketch the graph of depth (*d*) against time (*t*) for water being poured at a steady rate into these containers.

i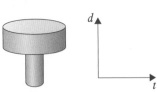

ii

How did you find this?   EASY   OK   HARD

SHAPE, SPACE AND MEASURES

55 Conventions, definitions and derived properties

Let's look at ...
● deciding between conventions, definitions and derived properties

✓ The sides of a triangle are labelled with the lower-case letter of the opposite angle.
This is called a **convention**.

✓ A **definition** is the minimum amount of information needed to specify a geometrical term.
Example   A polygon is a closed shape with straight sides.

✓ A **derived property** follows as a result of a definition.
It is not essential to a definition.
Example   The angles of a triangle add to 180°.

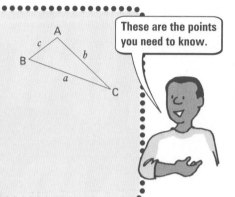

These are the points you need to know.

(A) Sort them out

Decide whether each of these is a convention (C), a definition (D) or a derived property (DP).

a   A square is a quadrilateral with all sides and all angles equal to 90°. _____

b   The diagonals of a square are equal and bisect each other at right angles. _____

c   We show lines are perpendicular using the symbol ⊥. _____

d   We name angles using the letter at the vertex or using three letters, the middle letter being the vertex. _____

e   A trapezium is a quadrilateral with one pair of parallel sides. _____

f   An equilateral triangle has three lines of symmetry. _____

g   A convex polygon is a polygon which has no reflex angles. _____

h   If two sides of a shape are equal, each side has a dash marked on it. _____

i   An image and its reflection in a mirror line have equal corresponding angles. _____

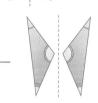

(B) Your turn

Look at each of these diagrams.
Complete the convention, the definition and the derived property for each.

a

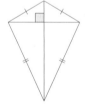

Convention:	We use two dashes on sides of a shape _____ _____
Definition:	A kite is a _____ _____
Derived property:	The diagonals of a kite _____ _____

b

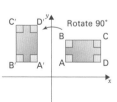

Rotate 90°

Convention:	If no direction is given for a rotation we rotate _____ _____
Definition:	A rectangle is a _____ _____
Derived property:	The diagonals of a rectangle _____ _____

How did you find this?   **EASY**   **OK**   **HARD**

56 Finding angles

Let's look at ...
● using geometrical reasoning to find unknown angles or prove something

These are the points you need to know.

✓ **Remember**

$a + b + c = 180°$
Angles on a straight line add to 180°.

$a = b$
$c = d$
Vertically opposite angles are equal.

$a + b + c + d = 360°$
Angles at a point add to 360°.

$a + b + c = 180°$
Angles in a triangle add to 180°.

$a = b$
Corresponding angles on parallel lines are equal.

$a = b$
Alternate angles on parallel lines are equal.

We often use **geometrical reasoning** to find an unknown angle or prove something.
Write down the steps clearly, one by one, and give reasons.

Example Prove that $x = 63°$

$$a = 180° - 126°$$ angles on a straight line add to 180°
$$a = 54°$$
$$x = b$$ base angles of isosceles △
$$x + b = 180° - 54°$$ angles of a triangle add to 180°
$$2x = 126°$$ because $x = b$
$$x = 63°$$

A Getting started

Find the value of the angles marked with letters.

a

$a =$ _____

b
$b =$ _____

c
$c =$ _____
$d =$ _____

d

$e =$ _____
$f =$ _____

e
$g =$ _____
$h =$ _____
$i =$ _____

f
$j =$ _____
$k =$ _____

B Prove these

Prove that x has the value shown. Show each step clearly and give reasons.

a

$x = 50°$

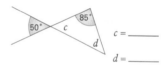

b ABCD is a trapezium. Work out the size of angle f. Show your working clearly.

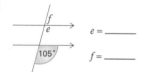

You may like to name some other angles. You could call this b.

57 More finding angles

Let's look at ...
● using geometrical reasoning to solve more difficult problems

✓ We can use **geometrical reasoning** to find unknowns or to prove something.

These are the points you need to know.

Example Find the value of **x**.
ABC is an isosceles triangle.
∠B = ∠C

$$2x + 10° = 3x - 4°$$ base angles of an isosceles triangle are equal
$$2x + 10° - 2x = 3x - 4° - 2x$$ subtracting $2x$ from both sides
$$10° + 4° = x - 4° + 4°$$ adding 4° to each side
$$x = 14°$$

You need an equation and a reason for each fact that you write down.

HELP
Finding this tricky? Make sure that you have completed sheet 56 **before** this page.

A Climbing parallelogram

a This diagram shows a parallelogram that just touches an equilateral triangle.
Find the size of angle *x*.
Show your working.

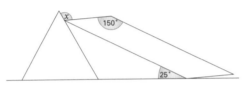

b Now the parallelogram just touches the equilateral triangle and ACDE is a straight line.
Show that triangle BCD is isosceles.

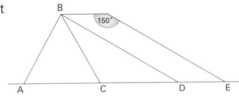

B Eight triangles

This pattern has rotation symmetry of order 8.
Find the size of angle *t*.
Show your working.

C Find *x*

Calculate the value of *x* by writing and solving an equation.

a

b

*D Tricky finish

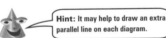

Hint: It may help to draw an extra parallel line on each diagram.

Find the value of the angles marked with letters. Show your working.

a

b

How did you find this? **EASY** **OK** **HARD**

58 Interior and exterior angles of a polygon

Let's look at ...
● calculating interior and exterior angles of regular and irregular polygons

These are the points you need to know.

✓ The **sum of the interior angles of a polygon with** n **sides is** $(n - 2) \times 180°$.

Example   This is a regular hexagon. It has **6** sides.
Sum of interior angles $= (6 - 2) \times 180° = 720°$
All interior angles in a regular hexagon are equal. Each interior angle is $\frac{720°}{6} = 120°$.

✓ **The sum of the exterior angles of any polygon is 360°.**

Example   Exterior angles add to 360°.
$x + 46° + 88° + 75° + 90° = 360°$
$x = 360° - 46° - 88° - 75° - 90°$
$= \textbf{61°}$

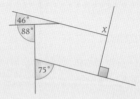

(A) Find a

Calculate the value of a.

a

$a =$ _____

b
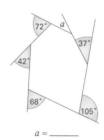
$a =$ _____

c
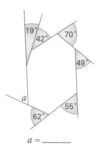
$a =$ _____

d
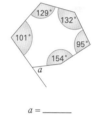
$a =$ _____

(B) Find the triangles

Any quadrilateral can be split into two triangles.

Explain how you know that the angles inside a **quadrilateral** add up to **360°**.

_____

_____

(C) Regular shapes

Complete this table.

Shape	Number of sides	Sum of interior angles	Size of one interior angle	Sum of exterior angles	Size of one exterior angle
Square	4	360°		360°	
Regular hexagon					
Regular octagon					
Regular 15–sided polygon					

*(D) Show your working

Find the value of x. Show your working.

a
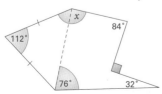

b
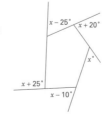

How did you find this?   EASY   OK   HARD

61

59 Triangles and quadrilaterals

Let's look at ...
● using the properties of 2-D shapes

These are the points you need to know.

✓ We use the **properties of 2-D shapes** to solve problems.

Example

isosceles trapezium

$a = 180° - 110°$ angles on a straight line add to 180°
$a = 70°$
$b = a$ an isosceles trapezium is symmetrical
$b = 70°$

A *True or false?*

Write true (T) or false (F) for each of these statements.

Remember: 'congruent' means 'exactly the same shape and size'.

a   A parallelogram has 2 lines of symmetry. ____

b   A rhombus has rotation symmetry of order 2. ____

c   A rectangle can be split into 4 isosceles triangles. ____

d   An isosceles trapezium can be split into 2 congruent triangles. ____

e   An arrowhead can be split into 2 congruent triangles. ____

B *What's missing?*

Use geometric reasoning and symmetry properties to find the angles marked with letters.

Show your reasoning clearly.

HELP

Need help with geometrical reasoning? See page 60.

a

rhombus

b
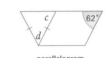
parallelogram

C *Why?*

Explain why all squares are rectangles. _____

_____

D *Imagine*

You will need extra paper.

a   Imagine you have four congruent equilateral triangles. You put the triangles together along sides of equal length. Draw and name the possible shapes that could be made.

b   Imagine you have a kite. You cut it in half to make two congruent triangles. Then you put the two triangles together along sides of equal length. Apart from a kite, what different shapes could you make?

*E Pythagoras' triangles

Pythagoras, a famous mathematician, discovered this formula was true for all right-angled triangles.

$$c^2 = a^2 + b^2$$

where c is the length of the longest side (hypotenuse), and a and b are the lengths of the other two sides.

Use Pythagoras' formula to find c. Round to 1 d.p. when necessary.

a

$c = $ ____

b

$c = $ ____

c
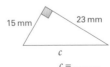
$c = $ ____

d
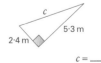
$c = $ ____

How did you find this?   **EASY**   **OK**   **HARD**

60 Tessellations

Let's look at ...
● tessellating regular and non-regular polygons

These are the points you need to know.

✓ A **tessellation** is made by reflecting, rotating or translating a shape.

Example

This tessellation could be made by rotating and translating the shape.

A Tessellate these

You may use more than one of these.

Tessellate each of these onto the grids.

Say which of reflection, rotation and translation you used.

The first one is started for you.

a
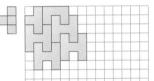

b
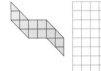

_____

_____

c

*d

_____

_____

B All regular

Equilateral triangle   Square   Pentagon   Hexagon   Heptagon   Octagon   Nonagon   Decagon

HELP

Need help with calculating internal angles? See page 61.

a i Three of the regular shapes above tessellate.
Write their names in the first column of the table, then calculate the size of one internal angle for each.

ii Use the table to help you explain why only some regular polygons tessellate. _____

_____

Regular polygons which tessellate	Size of one internal angle

*b There are eight tessellations that can be made using a combination of two or more regular polygons. Try tessellating these combinations. Each has been started for you.

i squares and equilateral triangles

ii octagons and squares
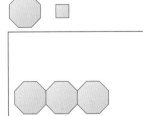

iii hexagons, squares and equilateral triangles

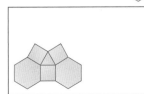

61 Circles

Let's look at ...
● naming the parts of a circle

These are the points you need to know.

✓ The **parts of a circle** are shown on this diagram.

✓ The **circumference** is the distance around the outside of the circle.

✓ A **tangent** is a line which just touches a circle at a point, P.

✓ When a line intersects the circle at two points, A and B, the line segment AB is a **chord**.

✓ A chord divides a circle into two **segments**.

✓ When a chord passes through the centre of a circle it is called a **diameter**. It divides the circle into two **semicircles**.

  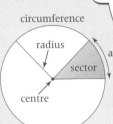

A Name that part

Name the parts of the circle shown in dark purple.

a _____
b _____
c _____
d _____
e _____

f _____
g _____
h _____
i _____
j _____

B Circle clues

Name the parts of a circle given by these definitions.

a A part of the circumference. _____

b A line segment joining any two points of the circumference. _____

c A region enclosed by a chord and an arc. _____

d A line outside the circle which touches a circle at just one point. _____

C Draw your own

 You will need compasses and a protractor and extra paper.

a Use compasses to draw a circle with radius 1·5 cm. Label these parts of your circle.

  i circumference
  ii sector
  iii diameter
  iv tangent

b Draw a regular pentagon by dividing the circumference of a circle into five equal arcs, using a protractor.

$\frac{360°}{5} = 72°$

How did you find this? **EASY** **OK** **HARD**

62 Constructions

Let's look at ...
- constructing triangles and quadrilaterals using a ruler and protractor

✓ We can **construct a triangle** using a ruler and protractor if we are given:
- two sides and the angle between them (SAS)

or
- two angles and the side between them (ASA).

These are the points you need to know.

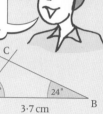

Examples

To construct this triangle:
1 Draw PR 2·6 cm long.
2 Draw an angle of 85° at R.
3 Draw RQ 2·8 cm long.
4 Join P to Q.

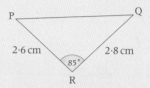

To construct this triangle:
1 Draw AB 3·7 cm long.
2 Draw an angle of 52° at A.
3 Draw an angle of 24° at B.
4 Label the intersection of these lines C.

(A) Triangle JKL

a Construct triangle JKL.

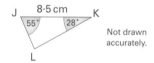

Not drawn accurately.

b Measure angle L on your diagram. _____

(B) Quadrilateral RSTU

a Construct quadrilateral RSTU.

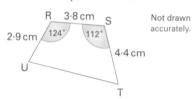

Not drawn accurately.

b Measure length TU. _____

(C) Hanworth Common

Three straight paths cut across Hanworth Common as shown.

Not drawn accurately.

a Complete the accurate scale drawing of the paths. Use the scale **1 mm represents 2 m**.

b Find the actual distance between the statue and the lake. _____

*(D) Is it possible?

You have a ruler and a protractor. Is it possible to construct triangle ABC such that

a A = 60°, B = 60°, C = 50° _____ If not why not? _____

b A = 40°, B = 70°, AB = 14 cm _____ If not why not? _____

How did you find this? **EASY** **OK** **HARD**

63 More constructions

Let's look at ...
● constructing triangles using a ruler and compasses

These are the points you need to know.

✓ If the three sides (SSS) of a triangle are given we can construct the triangle.

Example

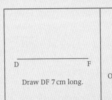

Draw DF 7 cm long.

Open compasses out to 4 cm. Draw an arc from D.

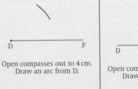

Open compasses out to 5 cm. Draw an arc from F.

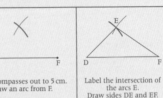

Label the intersection of the arcs E. Draw sides DE and EF.

✓ We can construct a right-angled triangle given the right angle, the length of the longest side (hypotenuse) and one other side (RHS).

Draw XZ 3 cm long.

Extend XZ. Construct a perpendicular at Z.

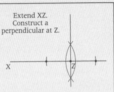

Open compasses out to 4 cm. With the point on X, draw an arc that crosses the perpendicular. Join X to the point of intersection. Label this point Y.

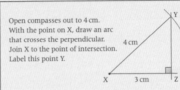

(A) Construct this

Use compasses to construct a triangle that has sides 6 cm, 5 cm and 3 cm.

Leave your construction lines.

(B) Sailmaker

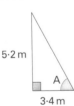

Pete sketches this diagram of the sail he needs for his boat.

a Make a scale drawing of the sail. Use the scale 1 cm represents 1 m.

b Use a ruler and protractor to find

　i the length of the longest side of the sail ____

　ii the size of angle A. ____

Measure angle A on your diagram, not the sketch.

(C) The big top

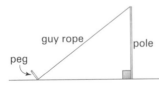

The ground crew are erecting a circus tent.

Each outside pole is 6 m tall.

Each guy rope is 10 m long.

a Make a scale drawing to show this information.

b How far from the pole will the guy rope be pegged into the ground? ____

c Find the angle between the guy rope and the ground. ____

　　　　How did you find this?　[EASY]　[OK]　[HARD]

64 Locus

Let's look at ...
● finding paths (loci) that follow rules or show real-life situations

✓ A **locus** is a set of points that satisfy a rule or a set of rules.

Examples The locus of the head of a boy jumping from a tree might look like this.
The locus of a robot moving so that it is always the same distance from a fixed point is a circle.

These are the points you need to know.

✓ We can **construct a locus**.

Example Your locus if you walked so that you were always the same distance from two trees would be the perpendicular bisector of the line joining the two trees.

tree ×------×tree

A Match it

a Match the loci of these with a sketch from the box.

> A ◯   B 〰️   C ⌒

  i A rugby ball kicked through the posts _____
  ii The pedal of an exercise bike _____
  iii An inflated balloon is let go _____

b Match these descriptions with loci from the box.

  i A goat is chained to a post. It moves so that it is always the same distance from the post. _____

  > **A** Perpendicular bisector of the line joining two points
  > **B** A circle
  > **C** Bisector of the angle between two lines

  ii A horse canters so that he is always the same distance from two fences that meet at right angles. _____
  iii Diana skis so that she is always the same distance from two trees. _____

B Graph it

a The locus of all points that are the same distance from (⁻2, 1) and (4, 1) is a straight line.
Draw this straight line. Label it with its equation.

b The locus of all points that are the same distance from the *x*-axis as they are from the *y*-axis is **two** straight lines.
Draw both straight lines.
Label these lines with their equations.

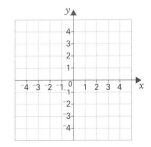

*C Sketch it

a Sketch the region in which point P would be if it is always closer to X than to Y.   X •   Y •

b Two doctors' surgeries are 300 m apart.

  i Doctor A takes patients from an area of radius 200 m around his surgery.
  Using the scale 1 cm represents 100 m, draw this area and shade it with horizontal lines. ≡

  ii Doctor B takes patients from an area of radius 150 m around his surgery.
  Draw this area and shade it with vertical lines. ‖‖‖

  iii Colour the area where patients can choose either Doctor A or Doctor B.

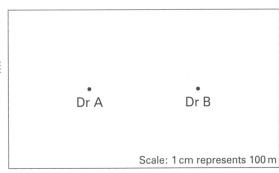

• Dr A   • Dr B

Scale: 1 cm represents 100 m

How did you find this? `EASY` `OK` `HARD`

65 3-D shapes

Let's look at ...
● representing 3-D shapes using isometric drawings, plans and elevations

✓ We can visualise or analyse **3-D shapes** from 2-D drawings, such as isometric drawings, or from cross-sections or descriptions.

✓ We can represent 3-D shapes using **plans and elevations**.
The view from the top of a shape is called the **plan view**.
The view from the front is called the **front elevation**.
The view from the side is called the **side elevation**.

These are the points you need to know.

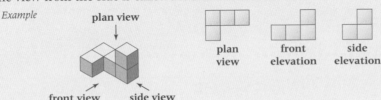

(A) Olivia's model

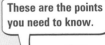

Olivia made this model with two purple cubes and six white cubes.
Olivia drew her model from three different views.
Which view does each drawing show?

i   ii   iii

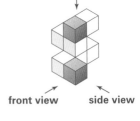

(B) Cameron's cubes

You will need some extra paper.

Cameron used six cubes to make a model.
These drawings show the front and side view of his model.

a Draw the plan, front and side elevations of this model.
  i Plan   **ii** Front   **iii** Side

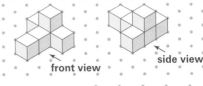

b

Cameron joins one more cube to his model.
This drawing from the front view shows where he joins the cube.
Complete the drawing from the side view.

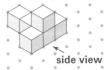

c Draw the plan, front and side elevations of this new model.
  i Plan   **ii** Front   **iii** Side

(C) Draw these

Draw three different shapes that have this plan.

plan

*(D) Sketch that solid

Sketch the solids that these describe.   You will need some extra paper.

a The front elevation is a square and the plan view and side elevation are both rectangles.

b The front and side elevations and the plan view are all circles.

c The plan view is a circle and the front and side elevations are triangles.

How did you find this?   EASY   OK   HARD

66 Cross-sections

Let's look at ...
● visualising and describing cross-sections
  of 3-D shapes

These are the points you need to know.

✓ When we slice a shape, the face that is made is called a **cross-section**.

Example If we slice a square-based pyramid parallel to the base as shown, the cross-section is a square.

base

(A) *Tetrahedron*

Felicity sliced a tetrahedron (triangle-based pyramid) horizontally near the top.

a What shape will the cross-section be? _____

b What if Felicity had sliced closer to the base?

_____

c What if Felicity had sliced closer to the top?

_____

(B) *Triangular prism*

James placed a triangular prism like this on the table.

a He sliced the triangular prism horizontally.

   i What shape was the cross-section? _____

   ii What if James had sliced closer to the top? _____

b What cross-section would James have got if he had sliced the triangular prism vertically? _____

(C) *Cylinder*

Miranda took this cylinder.
What shape would the cross-section be if she sliced.

a horizontally? _____

b vertically? _____

c diagonally? _____

(D) *Mark it*

a Use dashed lines to show where you would cut this cone to get an isosceles triangle cross-section.

b Use dashed lines to show where you would cut a cube to get these cross-sections.

   i square   **ii** rectangle   **iii** equilateral triangle   **iv** isosceles triangle

67 Congruence and transformations

Let's look at ...
- identifying congruent shapes, sides and angles
- reflections, rotations and transformations

> These are the points you need to know.

✓ **Congruent** shapes are exactly the same size and shape.
In congruent shapes
- sides in corresponding positions are equal
- angles in corresponding positions are equal.

✓ One congruent shape can be mapped onto
another by a translation, reflection or rotation
or some combination of these.

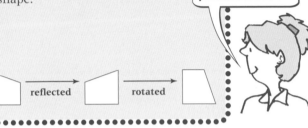

A Twisting triangles

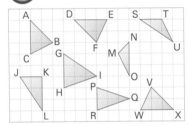

a Name the triangles which are congruent to triangle ABC. _____

b For each of the congruent triangles, name the side equal to
  i AC _____   **ii** BC. _____

c For each of the congruent triangles, name the angle equal to
  i B _____   **ii** C. _____

B Reflecting triangles

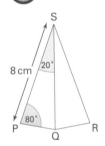

Triangle SQR is a reflection of triangle SQP in the line SQ. Fill in the gaps.

a **i** The length of SR is _____ cm.
  ii The size of ∠QSR is _____°.
  iii The size of ∠QRS is _____°.

b **i** △PSQ is isosceles because _____
  _____
  ii ∠SQR is _____°.
  iii The length of SQ is _____ cm.

C Transforming shapes

1 reflection in the *y*-axis		**2** reflection in the *x*-axis	
3 reflection in the line *x* = 1		**4** rotation of 90° about the origin	
5 rotation of 90° about (1, 1)		**6** rotation of 270° about the origin	
7 translation 4 units right and 5 units down		**8** translation 5 units right and 4 units down	

Shape A has been transformed in three different ways.

a Select the transformation from the box above which matches each diagram.

b For each transformation, decide whether A is or is not congruent to A'.

> Circle 'is' or 'is not'.

i
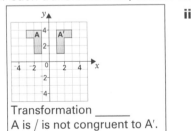
Transformation _____
A is / is not congruent to A'.

ii
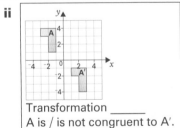
Transformation _____
A is / is not congruent to A'.

iii
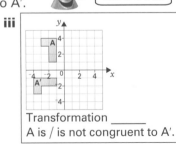
Transformation _____
A is / is not congruent to A'.

How did you find this?   EASY   OK   HARD

68 More congruence and transformations

Let's look at ...
- transforming 2-D shapes by reflection, rotation or translation
- the congruence of 2-D shapes before and after transformation

✓ When a shape is **reflected**, **rotated** or **translated**, the image is always **congruent** to the original shape.

Example ABCD is rotated 90° clockwise about the origin to A′B′C′D′. A′B′C′D′ is congruent to ABCD.

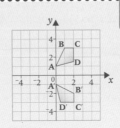

These are the points you need to know.

(A) Three transformations

a Transform ABCDE in each of the following ways. Write down the coordinates of A′B′C′D′E′ for each.

i

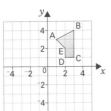

Reflect in the x-axis

A′ _____ B′ _____ C′ _____
D′ _____ E′ _____

ii

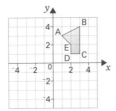

Translate 5 units left and 2 units down

A′ _____ B′ _____ C′ _____
D′ _____ E′ _____

iii

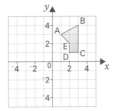

Rotate 180° about (0, 0)

A′ _____ B′ _____ C′ _____
D′ _____ E′ _____

b Are all of the image shapes found in **a** congruent to ABCDE? _____

c Is the image **always** congruent to the original shape when we
  i translate a shape? _____   **ii** rotate a shape? _____   **iii** reflect a shape? _____

(B) Transforming triangle

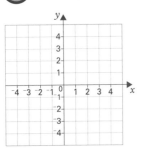

a The points (⁻2, ⁻2), (1, 2) and (4, ⁻1) are the vertices of a triangle. Plot these points on the grid.

b Write down the coordinates of the vertices after
  i translation 1 unit left and 2 units up _____
  ii reflection in the line x = 1 _____
  iii rotation 270° about the origin. _____

c Which of the images in **b** would be congruent to the original triangle you plotted? _____

(C) Rebecca's pattern

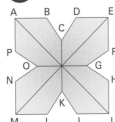

Rebecca made this pattern by reflecting and rotating this shape.

a Name three lines which are equal in length to each of these.

  i BC _____   **ii** MN _____

b Name angles which are equal to each of these.

  i ∠APO _____   **ii** ∠JKQ _____

69 Combinations of transformations

Let's look at ...
● transforming 2-D shapes by combining translations, rotations and reflections

These are the points you need to know.

✓ We can transform shapes using a **combination of transformations**.

Example A has been reflected in the *y*-axis and then translated 3 units down.

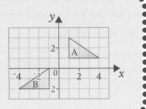

(A) *Imagine*

Imagine these transformations. What shape will the **combined** object and image form?

a A scalene triangle is reflected along one of its sides. _____ or _____

b A rectangle is rotated 180° about the mid-point of one of its sides. _____

c An equilateral triangle is rotated 60° about one of its corners. _____

(B) *Patrick Quinn*

Patrick designed this logo.
P maps onto Q using which combination?

A translation and enlargement

B translation and reflection

C translation and rotation.

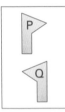

(C) *Transform this*

a Rotate triangle ABC 180° about the origin.
Label the image A′B′C′.

b Translate triangle A′B′C′ 6 units left and 2 units up.
Label the image A″B″C″.

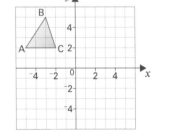

c Which single transformation maps △ABC onto △A″B″C″?
A translation   **B** enlargement
C reflection   **D** rotation

*d Describe the single transformation fully. _____

*(D) *Double trouble*

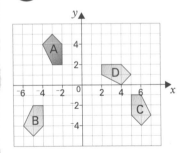

Shapes A, B, C and D are congruent.
Describe a combination of two transformations that will map A onto each of the other shapes.

a A → B 1. _____
           2. _____

b A → C 1. _____
           2. _____

c A → D 1. _____
           2. _____

How did you find this?   EASY   OK   HARD

70 Symmetry

Let's look at ...
- line symmetry, rotation symmetry and plane symmetry

Remember

These are the points you need to know.

✓ A shape has **line symmetry** or **reflection symmetry** if one or more lines can be found that divide the shape into two congruent shapes where one is the reflection of the other.

✓ A shape has **rotation symmetry** if it fits onto itself **more than once** during a complete turn.
The **order of rotation symmetry** is the number of times a shape looks identical in a complete turn.

We only count one of the start and finish positions.

✓ Some 3-D shapes are **symmetrical**.

Example The shaded shape is called a **plane of symmetry**.
It divides the shape into two congruent pieces such that one is a reflection of the other.
We say this shape has **plane symmetry**.

A 2-D shapes

Find the reflection and rotation symmetry of these.

a

- _____ lines of symmetry
- rotation symmetry of order _____

b

- _____ lines of symmetry
- rotation symmetry of order _____

c

- _____ lines of symmetry
- rotation symmetry of order _____

B Emma's rhombuses

Emma made this pattern out of rhombuses.
It has rotation symmetry of order 4 and 4 lines of symmetry.
Calculate the size of angles *a*, *b* and *c*. Show your working.

$a =$ _____ °

$b =$ _____ °

$c =$ _____ °

C Find the planes

a This shape has two planes of symmetry.
The first is shown. Draw the second.

b This shape has four planes of symmetry. Draw them all.

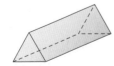

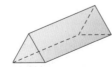

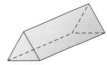

71 Enlargement

Let's look at ...
● using scale factor to enlarge a shape
● the effect of enlargement on lengths and angles

✓ To draw an enlargement you need to know the **scale factor** and the **centre of enlargement**.

Example   scale factor 2
centre of enlargement O

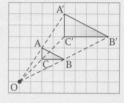

We label the image A'B'C'.
OA' = 2 × OA
OB' = 2 × OB
OC' = 2 × OC

✓ When a shape is enlarged, the ratio of any two corresponding line segments is equal to the scale factor.
So $\frac{A'B'}{AB} = 2$ $\frac{A'C'}{AC} = 2$ $\frac{B'C'}{BC} = 2$

These are the points you need to know.

✓ When a shape is enlarged, the shape and its image are **similar**. ABC and A'B'C' are similar.

✓ When a shape is enlarged, angles stay the same but lengths do not.

✓ If a shape with perimeter n is enlarged by scale factor **3**, the perimeter of the image is **3 × n**.

(A) Expanding arrow

This grid shows an arrow.
On the grid, draw an enlargement of the arrow of scale factor 2 .
Use point C as the centre of enlargement.

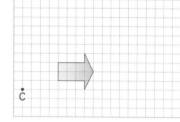

(B) Find the factor

In each of these diagrams the smaller shape and the larger shape are similar.
Use the dimensions given to find the scale factor of each of these enlargements.

The diagrams are drawn to scale.

a
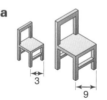

Scale factor _____

b

Scale factor _____

(C) Missing values

In each of these diagrams the smaller shape has been enlarged to the larger shape using the given scale factor. Find the missing values.

a

4 cm   101°   Scale factor 2   3·5 cm   x   A   y

$A =$ _____°, $x =$ _____ cm, $y =$ _____ cm

b

y   132°   Scale factor 1·5   10 cm   9 cm   A   x

$A =$ _____°, $x =$ _____ cm, $y =$ _____ cm

*(D) Puzzling perimeters

The perimeters of these shapes are given.
If each shape is enlarged by the given scale factor, what will the perimeter of the image be?

a
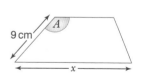
perimeter 5 m
scale factor 3

perimeter of image = _____

b
perimeter 15 cm
scale factor 5

perimeter of image = _____

How did you find this?   EASY   OK   HARD

72 Making scale drawings

Let's look at ...
● finding the scale of a scale drawing
● making a scale drawing

These are the points you need to know.

✓ A **scale drawing** represents something in real life. The real-life object is an **enlargement** of the object in the scale drawing.

✓ We can **find the scale** of a scale drawing.

Example In real life a building is 10 m high. On a scale drawing it is 5 cm high.
To find the scale we must find how many times 5 cm
has been 'enlarged' to give 10 m or 1000 cm.
The scale is **1 : 200** because $\frac{1000 \text{ cm}}{5 \text{ cm}} = 200$.

10 m is converted to 1000 cm so that the real-life height and the scale height have the same units

✓ We can **make a scale drawing** if we are given the real-life measurements and the scale.

Example A fence is 25 m long. To draw it using a scale of 1 : 500 we must work out how
long to draw 25 m on the drawing. A scale of 1 : 500 means that each cm on the
drawing will be 500 cm in real life.
Real-life fence length = 25 m = 2500 cm
So the scale drawing fence will be $\frac{2500}{500}$ = **5 cm** long.

─── 25 m ───

Ⓐ A school day

a In science, a scale drawing of a fish is 4 cm long. In real life the fish is 20 cm long.
What is the scale of the drawing?   1 : ____

b In geography, a map shows a 25 cm long river. In real life it is 25 km long.
   i   Convert 25 km to cm.   25 km = _____ cm
   ii   What is the scale on the map?   1 : ____

c In PE, a football field is 110 m long. A scale drawing shows it being 5·5 cm long.
   i   Convert 110 m to cm.   110 m = _____ cm
   ii   What is the scale on this drawing?   _____

Ⓑ Hamish's room

This sketch shows a plan view of Hamish's room. The figures on the sketch are in cm. The sketch is **not** drawn to scale. Make a scale drawing of this sketch. Use the scale **1 : 100**.

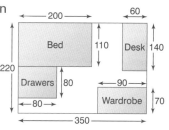

Ⓒ How long?

a A road is 500 m long. How long would the road be on a map if the scale was
   i   1 : 500? _____ m   **ii**   1 : 1000? _____ cm   **iii**   1 : 10 000? _____ cm

b A motorbike is 1·8 m long. How long would it be on a scale drawing with scale
   i   1 : 10? _____ cm   **ii**   1 : 20? _____ cm   **iii**   1 : 50? _____ cm

c The border between two countries is 150 km long. How long would this border be on a map with scale
   i   1 : 100 000? _____ cm   **ii**   1 : 1 000 000? _____ cm   **iii**   1 : 500 000? _____ cm

How did you find this?   **EASY**   **OK**   **HARD**

73 Interpreting scale drawings

Let's look at ...
● finding real-life measurements from scale drawings

✓ A **scale drawing** represents something in real life.
From the **scale** we can work out what each length in the drawing represents.

Example This is a scale drawing of a car.
Each millimetre on the drawing represents 6 cm in real life.
So 1 mm on the drawing represents 60 mm in real life.
The car is 51 mm on the drawing.
In real life it is 51 × 60 = 3060 mm
= 306 cm = 3·06 m.

Scale: 1 mm represents 6 cm

These are the points you need to know.

A At the zoo

Estimate the height of these.

a the deer _____

b the giraffe _____

c the tree _____

 Remember: An average man is about 1·8 m tall.

B Hay barn

This scale drawing shows a ladder leaning against a barn.

a How tall is the barn door in the drawing? _____

b How tall is the barn door in real life? _____

c How wide is the barn door in real life? _____

d What is the actual length of the ladder? _____

e What is the real-life angle that the ladder makes with the ground? _____

f A spider crawled up the left wall of the real barn, along the roof, and down the right wall of the barn. How far did it crawl altogether? _____

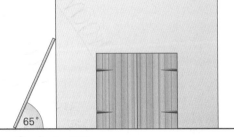

65°

Scale 1 : 100

*C Beachcomber Island

This map shows the island where Tagan and her family spend a week's holiday.

a i What is the map distance from the bird sanctuary to the waterfall?

_____

ii What is the actual distance from the bird sanctuary to the waterfall?
_____ km

c Find the actual distance from the airport to the shopping town. _____ km

d How far in metres, is the waterfall from Palm Beach? _____

e Tagan's family's hotel is at one of the three beaches. It is 11 km from the airport. Which beach are they staying at? _____

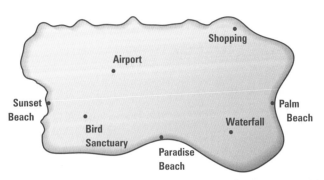

Shopping

Airport

Sunset Beach

Palm Beach

Waterfall

Bird Sanctuary

Paradise Beach

Scale 1 : 500 000

How did you find this? EASY   OK   HARD

74 Metric conversions, including area, volume and capacity

Let's look at ...
● converting length, mass, time, area, volume and capacity units

✓ **Remember:**

Length	1 km = 1000 m	1 m = 100 cm	1 m = 1000 mm	1 cm = 10 mm
Mass	1 tonne = 1000 kg	1 kg = 1000 g		
Capacity	1 ℓ = 1000 mℓ	1 ℓ = 100 cℓ	1 cℓ = 10 mℓ	1 ℓ = 1000 cm³	1 mℓ = 1 cm³
	1000 ℓ = 1 m³			
Area	1 hectare = 10 000 m²			
Time	1 hour = 60 minutes	1 week = 7 days		
	1 minute = 60 seconds	1 year = 365 days (366 in a leap year)		
	1 day = 24 hours	1 decade = 10 years		

These are the points you need to know.

✓ You need to know these **area, volume and capacity conversions**.

area
1 m² = 10 000 cm²
1 cm² = 100 mm²

capacity
1 cm³ = 1000 mm³
1 m³ = 1 000 000 cm³

(A) Your order please

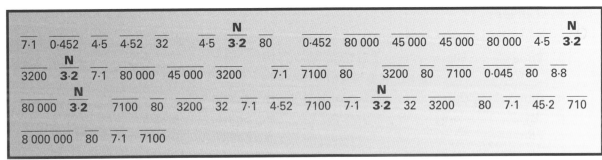

						N							**N**
7·1	0·452	4·5	4·52	32	4·5	**3·2**	80	0·452	80 000	45 000	45 000	80 000	4·5	**3·2**

N													
3200	**3·2**	7·1	80 000	45 000	3200	7·1	7100	80	3200	80	7100	0·045	80	8·8

	N								**N**					
80 000	**3·2**	7100	80	3200	32	7·1	4·52	7100	7·1	**3·2**	32	3200	80	7·1	45·2	710

| 8 000 000 | 80 | 7·1 | 7100 |

What goes in the gap? Write the letter beside each question above its answer in the box.

N 3200 mm = **3·2** m
V 45 mℓ = ____ ℓ
I 8 ha = ____ m²
U 4520 ℓ = ____ m³
L 4·5 m² = ____ cm²
H 710 000 mm³ = ____ cm³

O 4500 g = ____ kg
R 7·1 tonnes = ____ kg
D 88 000 m² = ____ ha
C 45·2 cm³ = ____ mℓ
E 8000 mm² = ____ cm²

T 0·32 m = ____ cm
A 71 mℓ = ____ cℓ
S 3·2 ℓ = ____ cm³
Y 8 m³ = ____ cm³
B 452 000 cm³ = ____ m³

(B) Scottish souvenirs

Gayle brought back this postcard and biscuit tin from her holiday in Scotland.

11 cm

17 cm

5 cm

15 cm   15 cm

Hint: Find **i** in each question, then convert this to find the other answers.

a What is the area of the postcard in
   i cm²? _____   **ii** mm²? _____   **iii** m²? _____

b What is the volume of the shortbread tin in
   i cm³? _____   **ii** mm³? _____   **iii** mℓ? _____   **iv** ℓ? _____

75 Working with measures

Let's look at ...
- solving measures problems
- working with metric and imperial equivalents
- calculating bearings and speeds

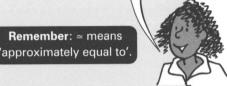

These are the points you need to know.

✓ These are some rough **metric and imperial equivalents**.

length	**mass**	**capacity**
5 miles ≈ 8 km	1 kg ≈ 2·2 lb	1 pint ≈ 600 mℓ
1 yard = 3 feet ≈ 1 m	1 oz ≈ 30 g	1 gallon ≈ 4·5 ℓ
1 inch ≈ 2·5 cm		1 litre ≈ 1·75 pints

Remember: ≈ means 'approximately equal to'.

✓ A direction from one place to another may be given as a **bearing**.

Bearings from North are always given as three digits.

To find the bearing of A from B:

Always draw the North line at the point you are measuring from.

1 Join AB.
2 Draw a North line at B.
3 Measure the angle in a **clockwise** direction between the North line and the line AB.

In this diagram the bearing of A from B is 075°. The bearing of B from A is 255°.

✓ **Average speed** can be calculated using this formula.

$$\text{average speed (km/h)} = \frac{\text{distance (km)}}{\text{time (h)}}$$

(A) Best prices

Circle the best price for each question. Explain how you worked out **a**.

a Rope   50p/m   or   20p/foot _____

_____

b	Ribbon	2p/cm	or	6p/inch	**c**	Taxi hire	80p/km	or	55p/mile
d	Steak	£5/kg	or	£2/pound	**e**	Saffron	£6·20/g	or	£175/oz
f	Fruit juice	£1·50/ℓ	or	95p/pint	**g**	Petrol	65p/ℓ	or	£2·80/gallon

(B) Madeleine's farm

You will need a protractor for this question.

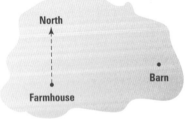

North

Barn

Farmhouse

This scale drawing shows the position of the farmhouse and the barn on Madeleine's farm. The scale is 1 : 1000

a How far apart are the farmhouse and the barn in real life? ____

b What is the bearing of the barn from the farmhouse? ____

c What is the bearing of the farmhouse from the barn? ____

(C) Fishing trip

Nathaniel spent a day fishing. First he drove to a river, and then he drove to a lake.

This diagram shows the distances and journey times for his trip.

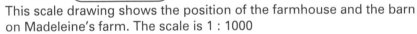

Home	River	Lake
	20 km	30 km
	15 mins	30 mins

a What was Nathaniel's average speed on his journey from home to the river? _____

b What was his average speed travelling from the river to the lake? _____

c What was his average speed over the whole journey from home to the lake? _____ Show your working.

How did you find this? [EASY] [OK] [HARD]

76 Perimeter and area

Let's look at ...
● calculating perimeters and areas
● solving problems with area and perimeter

These are the points you need to know.

Remember

✓ The distance right around the outside of a shape is called the **perimeter**.
Perimeter is measured in mm, cm, m or km.

✓ The amount of surface a shape covers is called the **area**.
Area is measured in mm², cm², m² or km².

Rectangle
Area = lb

Triangle
Area = $\frac{1}{2}bh$

Parallelogram
Area = bh

Trapezium
Area = $\frac{1}{2}(a+b) \times h$

✓ **Perimeters and areas** of complex shapes can be found by dividing them into simple areas then adding them.

Example   Area = area A + area B + area C
$= 8 \times 5 + 4 \times 3 + \frac{1}{2} \times 4 \times 4$ (length of dark purple line = 8 + 4 − 8 = 4 m)
$= 40 + 12 + 8$
$= 60 \text{ m}^2$

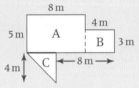

A Draw it

This rectangle has an area of 15 units².

Each square in the grid is 1 unit high and 1 unit wide.

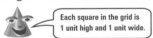

Draw these shapes on the grid.

a   A square with area 16 units²
b   A parallelogram with area 10 units²
c   An isosceles triangle with area 6 units²
d   A non-isosceles trapezium with area 7 units²
e   A rectangle with area 24 units² and perimeter 22 units

B Complex shapes

Calculate the area (A) and perimeter (P) of these shapes.

You will need some extra paper for your working.

a
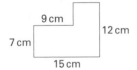
9 cm
12 cm
7 cm
15 cm
$A = $ _____ $P = $ _____

b
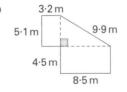
3·2 m
5·1 m   9·9 m
4·5 m
8·5 m
$A = $ _____ $P = $ _____

c
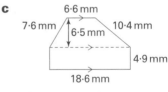
6·6 mm
7·6 mm   10·4 mm
6·5 mm
4·9 mm
18·6 mm
$A = $ _____ $P = $ _____

C Puzzles

a   The perimeter of square A is 64 cm.
The side length of square B is 64 cm.
The area of square C is 64 cm.

Use the information in the box to put squares A, B and C in order of size starting with the smallest. Order is ____ , ____ , ____ .

b   These two parks have the same area.

Remember: 1 hectare = 10 000 m²

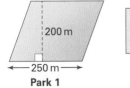
200 m
250 m
Park 1

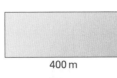
400 m
Park 2

i   How many hectares is park 1? _____
ii   What is the width of park 2? _____

How did you find this?   EASY   OK   HARD

77 Circumference

Let's look at ...
● using the formula for the circumference of a circle

These are the points you need to know.

 $C = \pi d$ or $C = 2\pi r$ give the **circumference**, C, **of a circle** of diameter d and radius r.
$\pi = 3 \cdot 14$ (2 d.p.) or you can use the π key on your calculator.

Example $C = 2\pi r$
$= 2 \times \pi \times 8$
$= \mathbf{50 \cdot 3}$ **mm (1 d.p.)** using the π key on a calculator

8 mm

A Quick questions

Find the circumference, C, of these circles using $\pi = 3 \cdot 14$.

a

10 mm

$C = $ _____

b

14 cm

$C = $ _____

c

4·5 m

$C = $ _____

Find the circumference, C, of these circles using the π key on your calculator.
Round your answers to 2 d.p.

d

13 cm

$C = $ _____

e

4·8 m

$C = $ _____

f

32·6 mm

$C = $ _____

B Harriet's house

a Harriet's circular dining table has a diameter of 1·8 m. What is the distance around the edge of the table? _____

b Harriet's circular coffee table has a circumference of 320 cm. What is its radius? _____

c The round mirror in Harriet's hallway has a radius of 25 cm. The round mirror in her bedroom has a diameter of 0·9 m. How much longer is the circumference of her bedroom mirror? _____

C Penny farthing

A penny farthing bicycle has one large wheel and one small wheel.

a The large wheel on a penny farthing has a diameter of 120 cm. Victoria rode it so that the large wheel went around exactly 5 times. How far did she travel? _____ m (2 d.p.)

b The small wheel has a diameter of 30 cm. How many times did it go around while the large wheel went around 5 times? ____

D Garden plots

Find the perimeter (P) of each of these garden plots.
Round your answers sensibly.

Each curve is either a semicircle or a quarter circle.

a

5 m
10 m

$P = $ _____

b

←—12 m—→

$P = $ _____

How did you find this? EASY   OK   HARD

78 Area of a circle

Let's look at ...
● using the formula for the area of a circle

These are the points you need to know.

✓ $A = \pi r^2$ gives the **area, A, of a circle** of radius r.

Example The area of this circle is
$A = \pi r^2$
$= \pi \times 12^2$
$= 452 \cdot 2 \text{ mm}^2$ (1 d.p.) using $\pi = 3 \cdot 14$

12 mm

(A) Quick questions

Use the π key on your calculator or $\pi = 3 \cdot 14$.

Remember: The formula uses radius not diameter.

Find the area (A) of each of these circles. Give the answers to 1 d.p.

a 7 m

b 9·1 cm

c 26 mm

d 12·4 m

$A = \underline{\hspace{2cm}}$ $A = \underline{\hspace{2cm}}$ $A = \underline{\hspace{2cm}}$ $A = \underline{\hspace{2cm}}$

(B) Willowbank Wildlife Park

Round your answers sensibly.

a A sprinkler sprays water in a circle of radius 4·8 m. What area of lawn is watered? _____

b A pony is tied to a post by a 6·2 m rope. What area of ground can the pony graze? _____

c The otter's lake is circular, with a diameter of 10·08 m. What is the area of the lake? _____

(C) Pizza parlour

Jason and Suna are dining at a pizza parlour.
They could buy one small pizza each, or share a large pizza.

a i Which choice will give them more pizza? _____
  ii How much more? _____

*b Laura orders one of the small pizzas on her own.
She eats a 270° sector of the pizza and has to leave the rest.
What area of pizza did she eat? _____

Choice 1   Choice 2

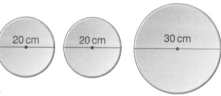

20 cm   20 cm   30 cm

(D) Shady shapes

*a i Find the area of the circle shown to the nearest m². ____
  ii The ratio of the area of the circle to the area of the square is 3 : 1. What is the length of the side of the square to the nearest m? ____

5 m

b Find the area (A) of each shaded shape.
All curves are semicircles or quarter-circles.

i   ← 15 mm →

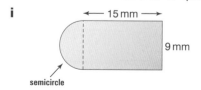

9 mm

semicircle

ii   ← 8 cm →

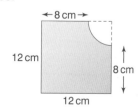

12 cm   8 cm

12 cm

$A = \underline{\hspace{1.5cm}}$ $A = \underline{\hspace{1.5cm}}$

How did you find this?   **EASY**   **OK**   **HARD**

79 Surface area and volume of a prism

Let's look at ...
● calculating the surface area and volume of a prism

These are the points you need to know.

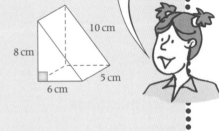

✓ A **prism** has a constant cross-section throughout its length.

✓ **Volume of a prism = area of cross-section × length**

Example   This is a triangular prism.
Volume of prism = area of cross-section × length
= area of triangle × length of prism
$= (\frac{1}{2} \times 8 \times 6) \times 5$

$= 24 \times 5$
$= \mathbf{120 \ cm^3}$

Surface area = sum of areas of all faces
= 2 × area of triangle + area base + area back + area sloping rectangle
= 2 × 24 + 6 × 5 + 8 × 5 + 10 × 5   area of $\triangle = \frac{1}{2} \times 8 \times 6$
= 48 + 30 + 40 + 50 $= \frac{1}{2} \times 48$
$= \mathbf{108 \ cm^2}$ $= \mathbf{24 \ cm^2}$

Ⓐ *Three prisms*

You will need some extra paper for all questions.

Remember the units.

The cross-section of these shapes is shaded.
Find the volume (*V*) of **a**, **b** and **c**. Find the surface area (*A*) of **a** and **b**.
Show your working.

a

4 mm   7 mm   11 mm

$V = \underline{\hspace{1cm}} \quad A = \underline{\hspace{1cm}}$

b

5 m   2·5 m   3 m   4 m

$V = \underline{\hspace{1cm}} \quad A = \underline{\hspace{1cm}}$

c
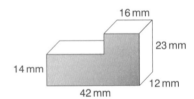
16 mm   23 mm   14 mm   42 mm   12 mm

$V = \underline{\hspace{1cm}}$

Ⓑ *Same size*

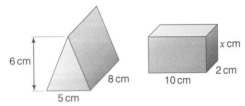
6 cm   8 cm   5 cm   *x* cm   10 cm   2 cm

These two shapes have the same volume.

a   What is the volume of the triangular prism? _____

b   What is the value of *x*? ____

Ⓒ *Skateboarders*

A skateboard ramp is in the shape of a prism.

a   The shaded end is a trapezium.
Calculate its area. _____

b   What volume of concrete was used to make the ramp? _____

c   The top and four sides of the ramp are to be painted.
What area will be painted? _____

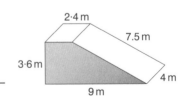
2·4 m   7·5 m   3·6 m   4 m   9 m

How did you find this?   EASY   OK   HARD

HANDLING DATA

80 Planning a survey

Let's look at ...
- writing conjectures
- deciding what data will be collected and the data source
- deciding sample size and degree of accuracy

✓ To **plan a survey**, these are the first three steps.

These are the points you need to know.

A Describe what **problem** you want to explore. To help you it is useful to formulate a **conjecture**.

Example 'Girls are better than boys at juggling three balls.'

> A conjecture is a statement you want to test the truth of.

B Decide **what data** needs to be collected and the possible sources of the data.
- You could use a **primary source**, for example
  - a survey of a sample of people
  - an experiment – observe, count or measure
- You could use a **secondary source**, for example websites, books, CD-ROMs, newspaper, ...

C Decide the **sample size** and the **degree of accuracy** for the data.
A sample should be as large as it is sensible to make it.

Example If you want to know if the height of the father affects the length of a newborn baby, you could measure the lengths of the babies to the nearest 0·5 cm, 1 cm or 2 cm.
You would need a sample size of at least 100 fathers and babies.

(A) Planning charts

Complete a planning chart for each of these two problems.

a

A PE teacher wants to know what factors affect the time it takes students to swim 50 m.
A Write two possible conjectures.   **1**   **2**

Choose one of your conjectures to continue planning with. Conjecture _____

B	**i**	What data will be collected?
	ii	What will the data source be?
	iii	Is this data source primary or secondary?
C	**i**	What will the sample size be and why?
	ii	What will the degree of accuracy be and why?

b

John wants to know what effect the wattage of a light bulb has on how many hours it lasts for.
A Write two possible conjectures.   **1**   **2**

Choose one of your conjectures to continue planning with. Conjecture _____

B	**i**	What data will be collected?
	ii	What will the data source be?
	iii	Is this data source primary or secondary?
C	**i**	What will the sample size be and why?
	ii	What will the degree of accuracy be and why?

How did you find this? **EASY** **OK** **HARD**

81 Collecting data

Let's look at ...
● designing a questionnaire

✓ Here are some **guidelines for writing a questionnaire**.

These are the points you need to know.

1 Allow for any possible answers.

Example   not at all ☐   rather than: up to 1 hour ☐
          up to 1 hour ☐   1 up to 2 hours ☐
          1 up to 2 hours ☐   2 up to 3 hours ☐
          2 up to 3 hours ☐
          more than 3 hours ☐

2 Give instructions on how you want the questions answered.   *Example*   Tick one of these boxes.

3 Do not ask for information that is not needed.   *Example*   Name of person.

4 If your questions are asking for opinions, word the questions so that *your* opinion is not evident.

5 Make the questions clear and concise.

It is a good idea to trial your questionnaire on a few people.

(A) *Is it true?*

A newspaper report says that 65% of people own a mobile phone.
Some students think that the percentage of people who own a mobile phone is actually higher.
They decide to do a survey.

a Dougal says 'I could ask 10 people from my Dad's work if they own a mobile phone.'
Give two different reasons why Dougal's method might not give very good data.

1 _____

2 _____

b Carey says 'I could hold up a sign at school at lunch time saying *"Please come and tell me if you own a mobile phone"*.'
Give two different reasons why Carey's method might not give very good data.

1 _____

2 _____

(B) *Fast food*

Which of these would be a better question to put in a questionnaire to find out how often people have takeaway meals?

A   Do you eat takeways?   **B**   Tick the number of times you usually eat takeaways in a month.
    Yes ☐   No ☐   0–2 ☐   3–5 ☐   6–8 ☐   ⩾9 ☐

Question _____ is better because _____

(C) *Fun at school?*

Louise's class was doing a survey on students' attitudes to school.
Rewrite each of these questions so that they would give more useful information.

a   I enjoy all of my school subjects.   Yes ☐   No ☐

b   I get to school on time.   Yes ☐   No ☐

c   I do my homework.   Yes ☐   No ☐

How did you find this?   **EASY**   **OK**   **HARD**

82 More collecting data

Let's look at ...
● designing a data collection sheet

These are the points you need to know.

✓ Many sets of data can be collected straight onto a **data collection sheet**.

✓ We group **continuous data** in equal class intervals.

Example

Height of plant (mm)	Tally	Frequency
$0 < h \leqslant 10$		
$10 < h \leqslant 20$		
$20 < h \leqslant 30$		
$30 < h \leqslant 40$		
$40 < h \leqslant 50$		

We usually have about 5 to 8 class intervals

← Class intervals must be equal width. They must not overlap.

A Science field trip

Class 9N go on a field trip to find the most common types of bird in a park.
Three students each suggest a data collection design.

Bridget's design	Callum's design	Dale's design
Use these codes to record the type of each bird that you see. Wren W Robin R Sparrow S Thrush T Example: R, R, T, R, S, S, W, ...	Write down the type of each bird that you see. Example: Robin, Robin, Thrush, Robin, Sparrow, Sparrow, Wren, ...	Use a tally chart to record the types of each bird you see. Example: <table><tr><td>**Type of bird**</td><td>**Tally**</td></tr><tr><td>Wren</td><td>I</td></tr><tr><td>Robin</td><td>III</td></tr><tr><td>Sparrow</td><td>II</td></tr><tr><td>Thrush</td><td>I</td></tr></table>

a Whose design should they **not** use? _____ Why? _____

_____

b Whose design is the best? _____ Why? _____

_____

B How long?

Carla designed this data collection sheet to collect data on the length of 13-year-olds' feet.

Name	Length of foot
Monica Church	23·2 cm

a Give two things about it that could be improved.

    1 _____

    2 _____

b Design a more suitable collection sheet.

Carla expects that the length of 13-year-olds' feet will be between 16 cm and 30 cm long.

C Design these

Use extra paper for this question.

Design a data collection sheet for these. Group the data sensibly.

a Heights of Year 9 boys

b Hours of sleep per night for adults

How did you find this?

83 Two-way tables

Let's look at ...
● interpreting and designing two-way tables

✓ A **two-way table** displays two sets of data.

Example This two-way table shows the ages of students in school sports teams.

	Netball	Football	Hockey
Under 14	16	25	27
14 to 16	36	53	26
Over 16	29	34	28

These are the points you need to know.

A *School of nations*

This table shows the place of birth of students at an international school.

	Junior school	Middle school	Upper school
Europe	57	35	43
America	12	31	21
Africa	7	7	12
Asia	15	18	26
Pacific	3	1	4

a How many students in total were born in Africa? _____

b How many students are in the middle school? _____

c How many upper school pupils were born in Asia or the Pacific? _____

d Does the junior, middle or upper school have the most students? _____ How many pupils is this? _____

e Compare the place of birth of junior and middle school students. _____
_____
_____

B *Fitness test*

Mr Chatfield recorded the number of laps run by his students in a fitness test.

Year 7

Number of laps	Girls %	Boys %
3	19	17
4	42	38
5	23	24
6	9	11
7	5	6

Year 8

Number of laps	Girls %	Boys %
3	15	11
4	36	32
5	27	29
6	8	11
7	6	6

Year 9

Number of laps	Girls %	Boys %
3	6	5
4	21	18
5	29	27
6	19	20
7	15	17

a Which year group had the greatest proportion of students running 7 laps? _____

b Compare the fitness test results of the three year groups and of boys and girls. _____
_____
_____

C *Natural and synthetic*

Mackenzie was studying the differences between wool and polarfleece.
She measured the drying time, stretch, durability and insulation factor of each.
Design a two-way table for the results.

84 Mode, median, mean

Let's look at ...
● **calculating and comparing the mean, median and mode for a set of data**

These are the points you need to know.

✓ **Range** = highest data value – lowest data value.

✓ The **mode** is the most commonly occurring data value.
The **modal class** is the class interval with the highest frequency.

✓ The **median** is the middle value of a set of ordered data.
Example The median of 14 g, 17 g, 19 g, 21 g, 22 g and 27 g is $\frac{19+21}{2} = 20$ g.

When there is an even number of values we find the mean of the two middle values to find the median

✓ **Mean** = $\frac{\text{sum of data values}}{\text{number of values}}$ The mean is affected by extreme values.

✓ Often one of the mean, median or mode represents the data the best.

(A) Gingernuts

These are the masses of ten packets of gingernut biscuits.
482 g, 520 g, 492 g, 513 g, 503 g, 496 g, 513 g, 509 g, 488 g, 504 g

a Find the mean, median and mode. Mean _____ median _____ mode _____

b Does the mean represent the data well? _____
Explain _____

Is the data evenly spread around the mean?

c Does the mode represent the data well? _____ Explain _____
_____

d This table shows the masses of 100 packets of gingernuts. What is the modal class for mass? _____

Mass	480–489	490–499	500–509	510–519	520–529
Frequency	8	19	24	43	16

(B) Parking tickets

The number of parking tickets given out by a warden each day for a week are 50, 48, 52, 47, 53, 0, 0.

a Find the mean, median and mode. Mean _____ median _____ mode _____

b Does the mean, median or mode represent this data the best? _____
Give a reason for your answer. _____

(C) First fifteen

The mean age of members of a rugby team is 13 years 4 months. The range is 2 years 0 months. Johnny, who is 14 years 4 months joins the team.

a What will happen to the mean age of the team? ☐
   A It will increase by exactly 1 year.   **B** It will increase by less than 1 year.
   C It will stay the same.   **D** It is not possible to tell.

b Which of **A**, **B**, **C** or **D** in part **a** is true about the range of ages in the team once Johnny has joined? ☐

(D) Three games

In a competition, each player plays three games. Find the three scores for each of the players.

a The mode of Lavinia's scores is 4. The mean of her scores is 6. Lavinia's scores are ___, ___, ___.

b The mean of Sam's scores is 8. The median and range are both 7. Sam's scores are ___, ___, ___.

c The range of Ben's scores is 6. The mean of his score is 5. Ben's scores could be ___, ___, ___ or ___, ___, ___.

How did you find this? **EASY** **OK** **HARD**

85 Displaying data on bar charts and frequency diagrams

Let's look at ...
- **understanding when to use bar charts and frequency diagrams**
- **displaying data on bar charts and frequency diagrams**

These are the points you need to know.

✓ A **bar chart** can be used to display categorical, discrete or grouped discrete data.

Categorical data is non-numerical data.

Discrete data can only have certain values.

✓ A **frequency diagram** is used to display continuous data.

Continuous data can have any values within a certain range

✓ More than one set of data can be displayed on **compound** bar charts and on frequency diagrams.

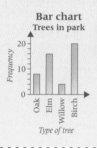
Bar chart
Trees in park

Bar charts can be drawn vertically of horizontally.

Compound bar chart
Winter show

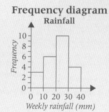
Frequency diagram
Rainfall

We label between the bars in a frequency diagram.

A Hockey All-stars

This table shows the results of the Hockey All-stars A, B, C and D teams over the last 4 years.

Team	Wins	Draws	Losses
A	40	10	20
B	16	8	54
C	28	5	42
D	62	7	6

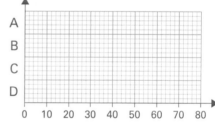

a Is this data categorical, discrete or continuous? _____

b Circle the type of graph which would display this data best.
- Bar chart   • Compound bar chart   • Frequency diagram
Explain why. _____

c Use the graph type which you selected in **b** to display this data. Remember to give your graph a title and label the axes.

d Which team **played** the most games over the last 4 years? _____

e Compare the performances of the A and B teams. _____

_____

B How fast?

The speeds of 100 vehicles on a stretch of motorway are given by this table.

Speed (mph)	50–	55–	60–	65–	70–	75–	80–85
Frequency	5	11	18	30	21	12	3

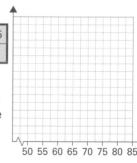

a Ian thinks that a frequency diagram is best to display this data. Explain why he is correct. _____

b Draw a frequency diagram for the data. Remember to give your graph a title and label the axes.

c How many of the vehicles were travelling at
  i less than 60 mph? _____   **ii** 70 mph or more? _____

d Into which class interval would you put a vehicle travelling at 69·9 mph? _____

How did you find this?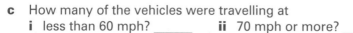

86 Displaying data on line graphs and pie charts

Let's look at ...
● understanding when to use line graphs and pie charts
● displaying data on line graphs and pie charts

✓ **Pie charts** can be used to display categorical or ungrouped discrete data.   *Example*

Eye Colour

There are 200 students at a school, 90 of them have blue eyes.
This is $\frac{90}{200}$ or $\frac{9}{20}$ of the students.
$\frac{9}{20} \times 360° = 162°$

These are the points you need to know.

✓ **Line graphs** are often used to show trends over time.
More than one set of data can be displayed on a line graph.

Price of bus ticket between two towns

A Asthma attacks

This table gives the proportion of cases of asthma per 100 000 people.

Your graph will need a title, a key and labelled axes.

Year	1981	1984	1987	1990	1993	1996	1999
Males	19·2	20·2	29·3	37·7	49·7	36·1	29·0
Females	16·7	16·8	25·2	36·9	51·0	40·2	34·8

a Display this data on a line graph. Put both sets of data on the same graph. Use ─●─ for males and ─■─ for females.

b In which year was the proportion of asthma cases highest for both males and females? _____

c Over which 3-year period did the proportion of asthma cases in females increase the most? _____

d Compare the asthma statistics for males and females. _____

_____

e Explain why a line graph will display this data best. _____

_____

B Heating survey

500 homes in a town are surveyed to find their main source of heating. This table gives the results.

Heating source	Wood	Gas	Electricity	Other
Frequency	75	210	122	93

a Draw a pie chart to show this data.
Write the angle at the centre of each sector.
Wood _____° Car _____° Electricity _____° Other _____°

*b In a second town, 900 homes are surveyed. When these results are graphed the sector for 'Wood' is exactly the same size as the first town. How many homes in the second town use wood as their main source of heating? _____

How did you find this?   **EASY**   **OK**   **HARD**

87 Scatter graphs

Let's look at ...
● interpreting and drawing scatter graphs and lines of best fit

These are the points you need to know.

✓ A **scatter graph** displays two sets of data.

✓ A scatter graph shows if there is a **correlation** (relationship) between the variables.

Example

Graph 1

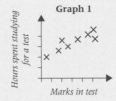

Hours spent studying for a test

Marks in test

Graph 2

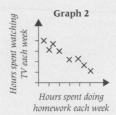

Hours spent watching TV each week

Hours spent doing homework each week

Graph 3

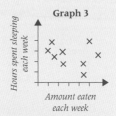

Hours spent sleeping each week

Amount eaten each week

Graph 1 shows there is a positive correlation between the hours spent studying for a test and the marks gained in the test. As the number of hours increases so do the marks in the test

Graph 2 shows a negative correlation between hours spent watching TV and hours spent doing homework. As the number of hours spent watching TV increases, the hours spent doing homework decreases

Graph 3 shows there is no correlation between the hours spent sleeping and the amount eaten in a week

*✓ If the data shows some correlation we can draw a **line of best fit**. There should be about the same number of points above the line as below it.

Ⓐ *The northern pike*

This scatter graph shows information about fish called northern pike.

a What does the graph show about the correlation between the head length and the body length of the northern pike? _____

b The body of a different fish is 125 cm long. Its head is 22 cm long. Use the graph to explain why this fish is **not** likely to be a northern pike. _____

_____

c Another fish **is** a northern pike. Its head length is 29 cm. Estimate its body length. _____

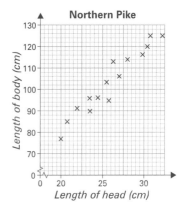

Northern Pike

Length of body (cm)

Length of head (cm)

Ⓑ *Football fanatics*

A football stadium has a problem with unruly fans during matches. This data shows the number of police present and the number of arrests made during ten home games.

Number of police	30	56	42	57	80	36	48	72	40	65
Number of arrests	52	37	43	26	7	40	32	11	7	22

a Draw a scatter graph of this data.
The scales have been started for you.

b Describe the correlation between the number of police and the number of arrests. _____

c One game was played on the day of a major train strike. Which point on the graph do you think shows this game? _____

*d Draw a line of best fit through **nine** of the points on your graph.

*e The stadium manager wants no more than 15 arrests made during the final game of the season. Approximately, what is the minimum number of police he should request? _____

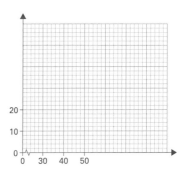

88 Interpreting graphs

Let's look at ...
● interpreting a variety of graphs

These are the points you need to know.

✔ We often use graphs to help us **interpret data**.

Example This graph represents the number of shells Morag found at two beaches.
From this graph we can tell
● Morag collected 50 shells altogether.
● Morag found more cockles than any other shell.
● Morag found more of each type of shell at Half Moon Bay than Pine Beach.

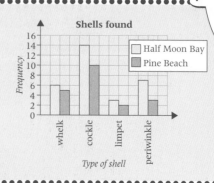

A How tall?

These graphs show the heights of students in two different classes at Avondale College.

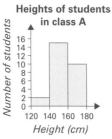

All classes at Avondale College have between 25 and 30 students in them.

a How many students are in class A?

26 ☐   27 ☐   28 ☐   29 ☐   not possible to tell ☐

b How many students are in class B?

26 ☐   27 ☐   28 ☐   29 ☐   not possible to tell ☐

c Which class is taller overall? _____ Explain how you know. _____

_____

B How old?

The organisers of a fun run print this graph showing the ages of runners.

a About what percentage of female runners were aged
   i 0–19 years? _____   **ii** 60–79 years? _____

b There were approximately 400 male runners aged 0–19.
   Estimate the number of male runners aged 20–39. _____

c The total number of male runners was about the same as the total number of female runners.
   Tick the correct statement.
   ● *Generally, the male runners were younger than the female runners.*   ☐
   ● *Generally, the female runners were younger than the male runners.*   ☐

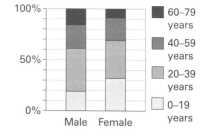

How did you find this?   EASY   OK   HARD

89 More interpreting graphs

Let's look at ...
● interpreting a variety of graphs

These are the points you need to know.

✓ We often use graphs to **interpret data**.
This page follows on from page 92. Make sure that you have completed page 92 before starting this page.

A Athletics day

This frequency diagram shows the long jump results for Oldwood School.
The jumps were measured to the nearest centimetre and classified in intervals 0 ≤ d < 50, 50 ≤ d < 100, ...

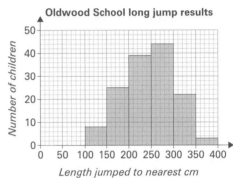

Oldwood School long jump results

Number of children (vertical axis)
Length jumped to nearest cm (horizontal axis)

a What is the modal class interval? _____

b Tamara jumped 150 cm. Which class interval was her jump put into? _____

c Jack jumped 320 cm. He said that he came fourth in the school. Could he be correct? _____

Explain. _____

*d Sophie jumped 251 cm. She said 'I was above the median.' Is she correct? _____ Explain how you know. _____

B Five plus a day

Rosie, a university student, was studying changes in the British diet over time. She graphed the average amounts of fresh potatoes, vegetables and fruit eaten.

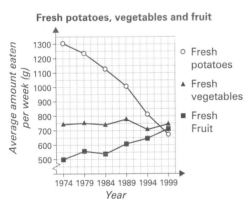

Fresh potatoes, vegetables and fruit

Average amount eaten per week (g) (vertical axis)
Year (horizontal axis)

○ Fresh potatoes
▲ Fresh vegetables
■ Fresh Fruit

a Complete this sentence. 'The amount of fresh _____ being eaten changed very little from 1974 to 1999.'

b Describe what has happened to the amount of fresh fruit eaten. _____

c Describe the change in the amount of fresh potatoes eaten. _____

_____

d Over which 5-year period did the consumption of fresh potatoes, vegetables and fruit **all** decrease?

_____

C Seal time

Two wildlife parks each have five species of seals. Sea World has 14 seals in total and Ocean Park has 37 altogether.

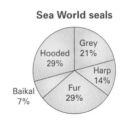

Sea World seals

Grey 21%
Hooded 29%
Harp 14%
Fur 29%
Baikal 7%

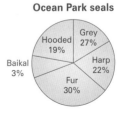

Ocean Park seals

Grey 27%
Hooded 19%
Harp 22%
Fur 30%
Baikal 3%

a The sum of the percentages in the Ocean Park pie chart is not 100%. Does this mean there must be a mistake? _____ Explain your answer. _____

b Zac said that Sea World has more hooded seals than Ocean Park. Is this true or false? _____ Explain. _____

c Which wildlife park has more harp seals? _____ How many more? _____

How did you find this? EASY OK HARD

93

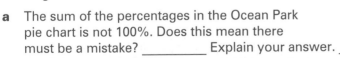

90 Misleading graphs

Let's look at ...
● identifying and describing misleading features of graphs

✓ Sometimes graphs, diagrams or statements can be **misleading**.

Example

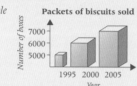

Packets of biscuits sold

This graph is misleading because the volume of the boxes makes it look like the increase has been very great.
Also, the vertical scale does not start at zero, again exaggerating the increase in the number sold.

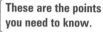
These are the points you need to know.

✓ To decide whether or not a graph is misleading, think about these questions.
● Is the graph complete?
● Is the scale appropriate?
● Are conclusions made from very small samples?
● Is the data graphed appropriately?

A Cheese profits

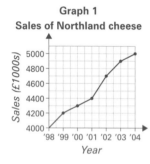

Graph 1
Sales of Northland cheese

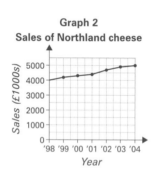

Graph 2
Sales of Northland cheese

a Which graph gives the impression of rapidly increasing sales? _____

b Have sales really been increasing rapidly from 1998 to 2004? _____

c If you were the owner of Northland cheese, which graph would you want to publish? _____ Why? _____

_____

_____

B Find the problem

Describe the misleading feature of each of these graphs.

a

Huge drop in train fares

_____

_____

b

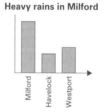

Heavy rains in Milford

_____

_____

c

Wool exports

_____

_____

d

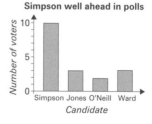

Simpson well ahead in polls

_____

_____

How did you find this? [EASY] [OK] [HARD]

91 Comparing data

Let's look at ...
● using the shape of distributions to compare data
● using the range and the mean, median or mode to compare data

✔ We can **compare data** by
  ● looking at the 'shape' of distributions

Example

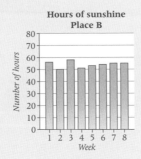

Hours of sunshine Place A

Hours of sunshine Place B

The two distributions have a different shape. For place A the number of hours of sunshine per week is much more variable than for place B.

> These are the points you need to know.

  ● using the range and either the mean, median or mode.

Example Robert found the mean and range of the hours of sunshine per week in two places.

Place A   mean 53   range 24
Place B   mean 54   range 8

The two places have about the same mean sunshine hours but place A has a much less consistent number of sunshine hours, shown by the bigger range.

A Cross country

The inter-school cross country competition is next week. Miss Walters wants to select either Jane or Hannah to represent the school.

These lists show the girls' times to the nearest second for their last 14 practice runs.

Jane   312   319   333   343   355   300   330
             308   358   321   343   324   313   335
Hannah   311   321   332   337   334   313   309
             322   331   333   336   318   323   339

a Complete a frequency table for each girl, then complete the bar chart for both girls.

Jane

Time (secs)	Tally	Frequency
300–309		
310–319		
320–329		
330–339		
340–349		
350–359		

Hannah

Time (secs)	Tally	Frequency

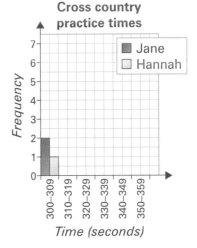

Cross country practice times
■ Jane
□ Hannah
Frequency
Time (seconds)

b Compare the shapes of the distributions for Jane and Hannah. _____
_____

c Find the mean, median and range for each girl's times.
  Jane   Mean _____   Median _____   Range _____
  Hannah   Mean _____   Median _____   Range _____

d Which girl would you select to represent the school? You could choose either as long as you justify your answer. _____
_____

92 Surveying

Let's look at ...
● planning and carrying out a survey

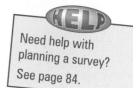

Need help with planning a survey? See page 84.

Follow the steps below to carry out your own 'mini' survey.

A Plan your survey

a What do you want to find out?

Some ideas are:
● How does time of day affect traffic patterns?
● How has immigration in Britain changed in the last 50 years?
● What factors affect people's opinions of texting?

b What are some related questions?

c What might you find out?
Make up a conjecture to test.

d What data do you need to collect?

e How accurate does the data need to be?

f What will the data source be?

It could be a **primary** source like asking people or experimenting, or a **secondary** source like books or the Internet.

g What will the sample size be?

www.statistics.gov.uk is a useful secondary source website.

B Collect your data

Design a questionnaire or data collection sheet.
Collect your data, and attach it to this page.

Need help with collecting data? See pages 85 and 86.

C Display your results

Use a suitable graph to display your results.
Draw this on some extra paper and attach it to this page.

Need help with displaying data on a graph? See pages 89, 90 and 91.

D Analyse your results

Find the mean, median, mode and range if appropriate.

E Write a report

Write a report on what you found out. Include your conclusions.
Make sure your conclusions relate to the original problem you set out to investigate.
In your report, write about any difficulties you had and how you solved them.
Suggest any improvements you could make next time.
Attach your report to this page.

How did you find this?   EASY   OK   HARD

93 Language of probability

Let's look at ...
● **using the language of probability to describe the likelihood of events**

These are the points you need to know.

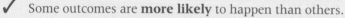

✓ We often **describe probability** using these words.
   certain, possible, impossible, 50% chance, even chance, better than even chance, less than even chance, very likely, likely, very unlikely, equally likely

✓ When an event is **random**, the outcome is **unpredictable**.

✓ Some events have **equally likely** outcomes.

   Example   Tim takes a marble out of this bag at random.
   It is **equally likely** to be a purple or white marble because there is the same number of purple and white marbles in the bag.

Always simplify fractions if you can.

✓ Some outcomes are **more likely** to happen than others.

   Example   Jody takes a marble out of each of these bags at random. She is **more likely** to take a purple marble from Bag A than Bag B because there is a greater proportion of purple marbles in Bag A than Bag B.
   Bag A: $\frac{3}{7} = 0.43$ (2 d.p.)   Bag B: $\frac{4}{13} = 0.31$ (2 d.p.)

A   B

A Do you agree?

For each of these statements say whether you agree or disagree, and why.

a   Mrs Rayward has four children, all four are boys. She is going to have another baby. She says, '*My next baby can't **possibly** be a girl*'. _____

b   The next person to buy a ticket to a football match could be a male or a female. So there is a **50% chance** that it will be a male. _____

c   Jordan buys a ticket for the chocolate wheel at his school fair. He says, '*I'm lucky, I'm **certain** to win*'. _____

B Mint or caramel?

I have two boxes of chocolates. Each chocolate looks the same on the outside.

   Box A contains 3 mints and 5 caramels.   Box B contains 3 mints and 9 caramels.

I am going to take one chocolate from either box A or box B.

I prefer mint chocolates. From which box am I more likely to get a mint? _____
Why? _____

*C Troll attack

In a computer game, trolls are hidden under circles. When you land randomly on a circle with a troll you are out of the game.

screen 1   screen 2   screen 3

a   The crosses indicate where the trolls are hidden on three different screens.
   On which of the three screens is it hardest to survive? _____

b   On which of these screens is it hardest to survive? _____
   Screen 4: 13 trolls out of 25 circles   **Screen 5**: 19 trolls out of 36 circles
   Screen 6: 31 trolls out of 60 circles

94 Mutually exclusive events

Let's look at ...
● deciding whether or not events are mutually exclusive

These are the points you need to know.

 Events that cannot happen at the same time are called **mutually exclusive**.

Example A is the event 'getting an even number' when a fair dice is rolled.
B is the event 'getting a 3' when a fair dice is rolled.
These events cannot happen at the same time.
They are mutually exclusive events.

A Toss, roll and spin

 Circle the correct answer.

Decide whether or not events A and B are mutually exclusive.

a A coin is tossed.
Event A: a head
Event B: a tail
Mutually exclusive   Not mutually exclusive

b A dice is rolled.
Event A: an even number
Event B: an odd number
Mutually exclusive   Not mutually exclusive

c A dice is rolled.
Event A: a number greater than 3
Event B: a number less than 5
Mutually exclusive   Not mutually exclusive

d Two coins are tossed.
Event A: two tails
Event B: one head, one tail
Mutually exclusive   Not mutually exclusive

e This spinner is spun.
Event A: a multiple of 5
Event B: a multiple of 10
Mutually exclusive
Not mutually exclusive

f This spinner is spun.
Event A: odd number
Event B: purple
Mutually exclusive
Not mutually exclusive

B People watching

Can they both be true?

Jody sees three people at a bus stop.

State whether events A and B are mutually exclusive or not.

a The teenager has blue eyes.
The teenager has black hair. _____

b The elderly man uses a walking stick.
The elderly man wears a hearing aid. _____

c The child has a birthday in May.
The child has a birthday in the second half of the year. _____

C Card games

Sean has these eight cards. He selects two at random.
Circle the two mutually exclusive events in each row.

a ● a heart   ● an odd number   ● a number less than 4

b ● a square   ● a circle   ● an even number

c ● a triangle   ● a prime number   ● a number greater than 8

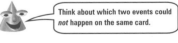 Think about which two events could *not* happen on the same card.

D Find your own

a Write one pair of events which **are** mutually exclusive and one pair which are not mutually exclusive.

 Be original!

A _____
B _____

A _____
B _____

How did you find this?   EASY   OK   HARD

95 Calculating probabilities of mutually exclusive events

Let's look at ...
● solving probability problems

✔ If the probability of an event occurring is p, then the **probability** of it **not** occurring is $1 - p$.

Example If the probability of Sophie getting a goal in netball is $\frac{5}{7}$, then the probability of her not getting the shot is $\frac{2}{7}$.

✔ **The sum of the probabilities of all the mutually exclusive outcomes of an event is 1.**

Example There are four different colours of counters in a bag, red, green, blue and yellow. One is taken at random.
The probabilities of getting three of the colours are
   red 0·2   green 0·25   blue 0·4
The sum of all four probabilities must equal 1.
So p(yellow) $= 1 - 0·2 - 0·25 - 0·4 = 0·15$

> These are the points you need to know.

A Shopping day

Simplify fractions when possible.

1 Margaret buys 10 nectarines, 4 peaches, 3 plums and 3 apricots. She puts them all in one bag. Margaret takes one piece of fruit at random from the bag. What is the probability of getting

   i a plum _____   **ii** a nectarine _____   **iii** not a peach? _____

2 Tony buys 16 muffins to share at a family gathering. The muffins come in 4 flavours. If a muffin is chosen from the box at random the probabilities of getting three of the flavours are
   chocolate $\frac{5}{16}$ blueberry $\frac{1}{4}$ banana $\frac{1}{8}$

   a What is the probability of getting the fourth flavour, bran? ____
   b Use the probabilities to work out how many muffins of each flavour were in the box.
      i chocolate _____   **ii** blueberry _____   **iii** banana _____   **iv** bran _____

3 A shop has purple, orange and blue kites in its window. There is at least one of each colour. If one is taken at random, the probability it is purple is $\frac{2}{3}$. There are 24 kites in the window. What is the greatest number of orange kites there could be? _____

B Sail away

An electrical store runs a competition to win a two week Greek island cruise. Each customer is entered into the draw.

This table gives the probability of selecting each age group.

Age (years)	Male	Female
0–19	0·1	0·05
20–39	0·2	0·1
40–59	0·15	0·15
60+	0·2	0·05

a What is the probability that the winning customer will be
   i less than 20 years old _____   **ii** a female? _____

b There were 1200 customers entered into the draw. How many of these were aged 60 or above? _____

*C How many leprechauns?

In an arcade game, a magical creature appears every time a screen is completed.
The probability of each creature occurring is   goblin $\frac{1}{5}$ elf $\frac{1}{3}$ centaur **?**   leprechaun $\frac{1}{15}$

a What is the probability of getting a centaur? _____

b Which creature is most likely to appear? _____

c What is the probability of not getting a leprechaun? _____

*d After many games, Nathan had completed many screens and seen 4 leprechauns appear. About how many screens do you think Nathan had completed? _____

How did you find this?   EASY   OK   HARD

96 Listing mutually exclusive outcomes

Let's look at ...
● listing outcomes as a sample space

These are the points you need to know.

✓ Before we can calculate probability we need to identify all the **mutually exclusive outcomes**.

The set of all possible outcomes is called the sample space.

Example A fair coin and a fair dice are thrown.
The possible outcomes can be shown in a table or a list.

Table:

Coin	Head	Head	Head	Head	Head	Head	Tail	Tail	Tail	Tail	Tail	Tail
Dice	1	2	3	4	5	6	1	2	3	4	5	6

List: H1, H2, H3, H4, H5, H6, T1, T2, T3, T4, T5, T6

'H' and 'T' stand for 'Heads' and 'Tails'.

A Shapely spinners

These two spinners are spun at the same time.

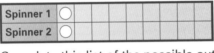

Spinner 1   Spinner 2

a Complete this table to show all the possible outcomes.

Spinner 1	○							
Spinner 2	○							

b Complete this list of the possible outcomes.
○ ○, _____

B Heads or tails?

a Two coins are tossed at the same time. There are four possible outcomes. List them.
_____, _____, _____, _____

b Three coins are tossed at the same time. List the eight possible outcomes.
_____, _____, _____, _____, _____, _____, _____, _____

C Add it up

Two fair dice are thrown at the same time and the numbers are added.

a Complete this table to show all the possible outcomes.

b What is the most likely total? _____
Why? _____

+	1	2	3	4	5	6
1	2	3	4			
2	3	4				
3						
4						
5						
6						

D Party in the Park

Angela, **B**radley, **C**arla, **D**aniel and **E**lgar are friends.

a They have three tickets to a concert.
  i Using the initials A, B, C, D and E list all possible groups of three who could go to the concert.
  _____
  ii How many possible outcomes is this? _____

b Elgar is given a fourth ticket. Now Elgar must go to the concert and three of the other friends will come with him.
  i List all possible groups of four who could now go to the concert. _____
  ii How many possible outcomes is this? _____

How did you find this? **EASY** **OK** **HARD**

97 Calculating probability by listing all the mutually exclusive outcomes

Let's look at ...
● identifying the sample space and using this to calculate probabilities

These are the points you need to know.

✓ To **calculate a probability** we often record all the possible **outcomes** using a list, diagram or table. The set of all the possible outcomes is called the **sample space**.

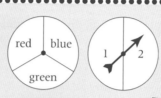

Example These two spinners are spun.
The possible outcomes are
red 1, red 2, blue 1, blue 2, green 1, green 2.

The probability of getting an outcome with blue $= \frac{2}{6}$ ◄── ways of getting blue
◄── total number of possible outcomes
$= \frac{1}{3}$

HELP

Need help with recording all of the possible outcomes?

Page 100 should be completed **before** this page.

Ⓐ *Night or day?*

These two spinners are spun at the same time.

a Complete this list of possible outcomes.
sun/sun,

b What is the probability of getting

　i two moons ____　**ii** a sun and a moon ____　**iii** at least one sun? ____

Ⓑ *Loose change*

Stephanie collects all of her loose 5p, 10p, 20p and 50p coins in a large money box.

There are equal numbers of each coin in the money box.

She randomly shakes two coins out of the box, into her hand.

a Complete this sample space to show how much money she could have in her hand.

	5p	10p	20p	50p
5p	10p	15p	25p	
10p				
20p				
50p				

b Use the sample space to calculate the probability of getting

　i a total of £1 _____
　ii a total less than 50p _____
　iii a total more than 35p. _____

Ⓒ *Glittering frames*

Natasha is making a photo frame for her mother's birthday.

She needs to choose a colour for the wood, the border and the background.

These are the colours she has.

Wood	**Border**	**Background**
Gold (G)	Red (R)	Cream (C)
Silver (S)	Blue (B)	White (W)

a List all the possible combinations Natasha could choose.

_____

b Natasha decides to choose the three colours randomly.
What is the probability that the frame

　i is gold, red and white _____　　**ii** has silver and white on it _____
　iii has neither gold nor blue on it _____　　**iv** has red or cream, but not gold? _____

How did you find this?　[**EASY**]　[**OK**]　[**HARD**]

98 Estimating probabilities from relative frequency

Let's look at ...
- finding relative frequencies
- using the relative frequency as an estimate of probability

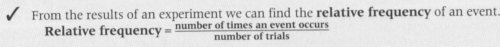

These are the points you need to know.

✓ From the results of an experiment we can find the **relative frequency** of an event.

$$\text{Relative frequency} = \frac{\text{number of times an event occurs}}{\text{number of trials}}$$

✓ We often use the relative frequency as an **estimate of probability**.

Example A student dropped a piece of toast with honey on it 50 times.
It landed honey side up 19 times.
The relative frequency of it landing honey side up is $\frac{19}{50}$.
We can use this to estimate that the probability of this toast landing honey side up is $\frac{19}{50}$.

✓ It is important to note that:

1 when an experiment is repeated, there may be, and usually will be different outcomes
2 increasing the number of times an experiment is repeated usually leads to better estimates of probability.

Ⓐ *Tin can surprise*

The Mitchell family bought hundreds of tin cans at an exceptionally cheap price: however, none of the cans had labels on them.

For the first month, Davey Mitchell recorded the contents of the cans as they were opened.

Contents	Peaches	Baked beans	Beetroot	Mangoes
Frequency	20	12	13	15

a How many cans were opened in the first month? _____

b Find the relative frequency of opening a tin of   **i** peaches _____   **ii** mangoes _____

c Estimate the probability of finding beetroot inside the next tin to be opened. _____

d If 200 cans were opened, how many times would you expect to find baked beans? _____

Ⓑ *Sparkling teeth*

Alex surveyed 200 randomly selected people to find out the brand of toothpaste they use. This table shows the results.

Brand	Frequency
Shine	19
So White	40
Dazzle	32
Minto	15
Budget	94

a Find the relative frequency of people who use

   i So white _____   **ii** Minto _____

b **i** Estimate the probability that the next person Alex surveys will use Budget brand. _____

   ii How could Alex improve the accuracy of this estimate? _____

c If Alex surveyed 500 people, how many would you expect to use Dazzle? _____

*Ⓒ *Guessing games*

a Sven has a guessing game on his computer. He won 100 of the games he played, so he estimated the probability of winning a game to be 0·5.
How many games did Sven play? _____

b Pip played the same guessing game. She won 6 of the games she played, and so she estimated the probability of winning each game to be 0·3.
How many games did Pip play? _____

c Whose estimate is likely to be better? _____ Explain why. _____
_____

How did you find this?   **EASY**   **OK**   **HARD**

99 Comparing experimental and theoretical probability

Let's look at ...
- comparing the results from a probability experiment with the theoretical probabilities
- deciding whether or not a game is fair

These are the points you need to know.

✓ When we compare **experimental probability** with **theoretical probability**, the greater the number of trials in the experiment, the closer the experimental probability is to the theoretical probability.

✓ Some games are **fair** and others are **unfair**.
If a game is fair it means that all players have the same chance of winning.

Example Two purple (P) and two white (W) counters are put in a bag.
Two counters are pulled out randomly.
One player gets a point if the counters are the same colour.
The other player gets a point if the counters are different colours.
To decide if the game is fair we list the sample space of the possible outcomes:
  PP, PW, WP, WW
There are two outcomes where the counters are the same colour (PP and WW) and two outcomes where they are different colours (PW and WP). So the game **is** fair.

A Games time

a Neil spun this spinner 20 times. He got red 11 times.

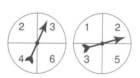

  i What is the theoretical probability of getting red? _____
  ii How many times would Neil expect to get red out of 20 spins? _____
  iii Neil thinks that there is something wrong with the spinner. Do you agree? _____
  Why or why not? _____

b The rules on a game of chance claim that the probability of winning each game is 0·85. Debbie plays this game 200 times and wins 166 times. She says, 'The rules must be wrong'. Do you agree with Debbie? _____ Why or why not? _____
_____

B Two spinners

Four students play a game using these two spinners.
Each student chooses a different rule card.

Hilary's Rule	Bernard's Rule	Kelly's Rule	Finbar's Rule
The total is even	The total is odd	The total is a square number	The total is a prime number

The two spinners are spun and the numbers added together. If the total satisfies a student's rule, they get a point.

a Draw a sample space of possible outcomes in the grid.

b Who is most likely to win? _____ Explain why. _____
_____

c Write the students' names in order from most likely to win to least likely to win.

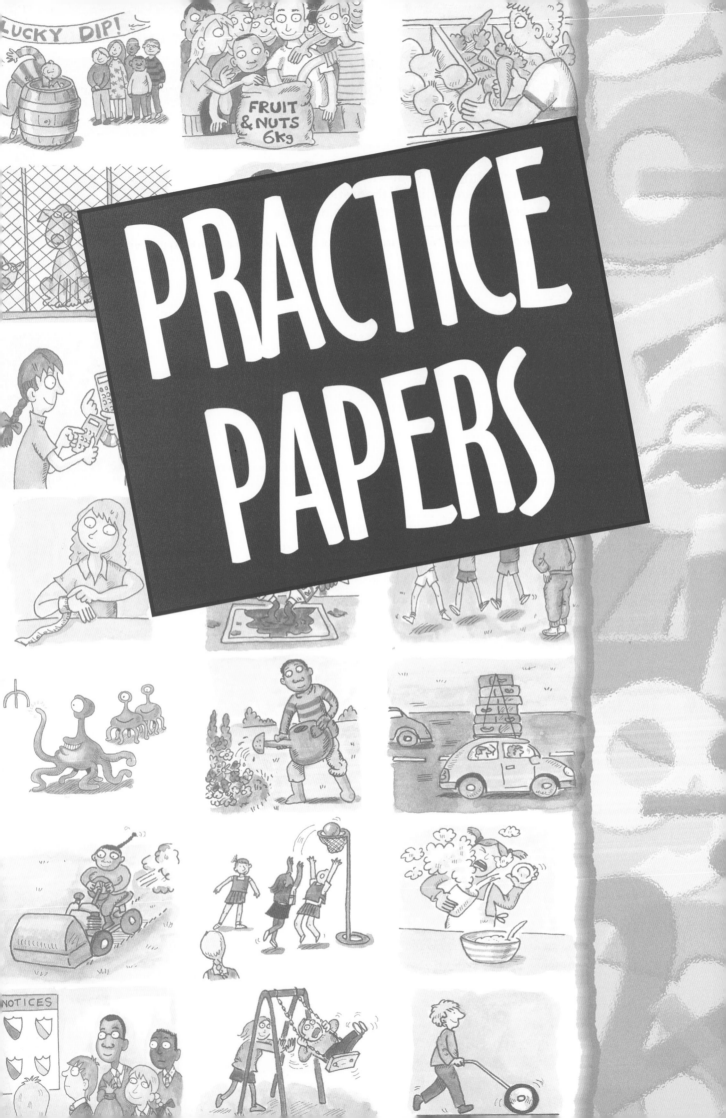

PRACTICE PAPERS

Paper 1

① Look closely

[Level 5]

Look at this diagram

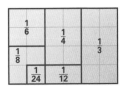

Use the diagram to calculate these. Make sure that all of your answers are written in their simplest form.

a $\frac{1}{4} + \frac{1}{12} =$ _____ **b** $\frac{1}{8} + \frac{1}{24} =$ _____ **c** $\frac{1}{3} - \frac{1}{6} =$ _____ **d** $\frac{1}{3} - \frac{1}{24} =$ _____

② Points of intersection

[SATs question Level 5]

The diagram shows two straight lines.
Where the lines **cross** is called a **point of intersection**.

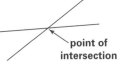

a Draw **three** straight lines that have only **one** point of intersection.

b **Three** straight lines have exactly **two** points of intersection.
Complete the sentence below.
Two of the lines must be _____ .

③ Daylight hours

[SATs question Level 5]

The graph shows at what **time** the sun rises and sets in the American town of Anchorage.
The day with the **most** hours of daylight is called the longest day.
Fill in the gaps below, using the information from the graph.

a The **longest day** is in the month of _____ .
On this day there are about _____ hours of daylight.

b The **shortest day** is in the month of _____ .
On this day there are about _____ hours of daylight.

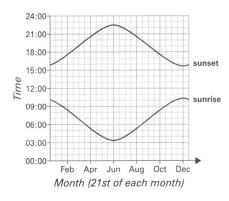

④ Best wishes

[Level 5]

Jeff buys a box of 40 greeting cards.
The cards come in these four different designs.

10 cards   12 cards   17 cards   1 card

Jeff takes a card from the box without looking.
What is the probability that the picture on the card is

a a dog _____ **b** a cat _____ **c** **not** a cat? _____

Paper 1 (cont.)

5 *Are they real?* [Level 5]

Susan buys a pair of gold painted earrings for £2·25.
A pair of real gold earrings would have cost her 18 times as much.

a How much do the real gold earrings cost? _____
Show your working.

b How much did Susan save by buying the gold painted earrings instead of the real gold earrings?
_____
Show your working.

6 *Hole in one* [Level 5]

Putt Land is a mini golf course. This graph shows their charges.

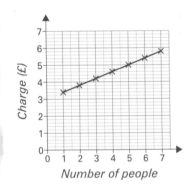

Number of people

a How much do *Putt Land* charge for
i 2 people _____ **ii** 3 people? _____

b For each extra person, how much more do *Putt Land* charge?
_____

c *Crazy Putt* is another mini golf course.
They charge £1 per person.
Draw a line on the graph to show this information.

d Complete this sentence.
Playing mini golf at *Crazy Putt* is cheaper if you have less than
_____ people.

7 *Is it magic?* [Level 5]

Mrs Knox showed her class a way to make a magic square. She said that you substitute numbers for x, y and z into these expressions.

$x-y$	$x+y-z$	$x+z$
$x+y+z$	x	$x-y-z$
$x-z$	$x-y+z$	$x+y$

a Judy made a magic square using the values $x = 9$, $y = 1$ and $z = 3$.
Complete her magic square.

8		
13		
6		

b Luke made this magic square.
What values did Luke use?
$x =$ _____ , $y =$ _____ , $z =$ _____ .

7	19	16
23	14	5
12	9	21

Paper 1 (cont.)

8 *Finish the functions* **[Level 6]**

a Each of these tables uses a function to map the number *n* to a different number.
Fill in the missing numbers.

i

n	→	$\frac{n}{3}$
24	→	8
15	→	
	→	6

ii

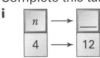

n	→	$n - 6$
9	→	
23	→	
	→	11

b Complete this table in two different ways.

i

| n | → | ___ |
| 4 | → | 12 |

ii

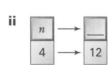

| n | → | ___ |
| 4 | → | 12 |

9 *Building bricks* **[Level 6]**

Tessa glues wooden cubes together to make bricks.
If she has 20 cubes, she can make one of the following four bricks.

	Brick A	Brick B	Brick C	Brick D
Length	20	10	5	5
Width	1	2	4	2
Height	1	1	1	2

a Which brick has the largest volume?

Brick A ☐   Brick B ☐   Brick C ☐   Brick D ☐   All the same ☐

b Which brick has the largest surface area?

Brick A ☐   Brick B ☐   Brick C ☐   Brick D ☐   All the same ☐

c How many of brick D would Tessa need to make a brick with length 10, width 4 and height 4?

_____

d If Tessa has 30 cubes, she can make one of five different bricks.

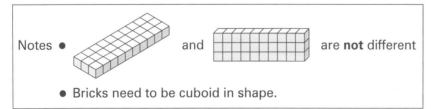

Notes ● and are **not** different

● Bricks need to be cuboid in shape.

Complete this table to show the length, width and height of the five different bricks.

	Brick 1	Brick 2	Brick 3	Brick 4	Brick 5
Length	10				
Width	3				
Height	1				

Paper 1 (cont.)

(10) *Sum and product* [Level 6]

Complete this table. The first column is done for you.

A	4	6	⁻3	
B	7	⁻2		
Sum of A and B	11		⁻7	
Product of A and B	28			

(11) *How many pines?* [Level 6]

Three-quarters of the trees in a park are evergreen.
Two-thirds of these evergreen trees are pine trees.
What fraction of the trees in the park are pines? _____
Show your working.

(12) *Rearranging*

a Rearrange the equations. [Level 6]

i $y - 2 = x$ $y =$ _____
ii $3f = e$ $f =$ _____
iii $r + 5 = 6q$ $r =$ _____

b Rearrange the equation to make n the subject. [Level 7]
Show your working.

$4(n - 3) = p$

$n =$ _____

(13) *Which road is quicker?* [Level 7]

There are two different roads which connect Dominic's town to his father's town. One is a motorway
and one is a country lane.
Each road takes Dominic the same amount of time.

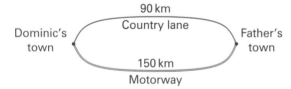

90 km
Country lane
Dominic's town
Father's town
150 km
Motorway

Dominic's average speed on the country lane is 60 km/h.
What is his average speed on the motorway?
Show your working.

_____ km/h

Paper 2

1 Shape rotation

[SATs question Level 5]

Look at this shape made from six cubes.
Four cubes are white.
Two cubes are purple.

Part of the shape is rotated through 90° to
make this shape.

After another rotation of 90°, the shape is a cuboid.
Draw this cuboid on the grid.

2 Salt Lake City

[Level 5]

This graph gives information about average
temperatures in Salt Lake City, USA.
For example, the average minimum temperature in
April is 0 °C.

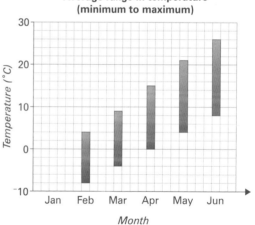

**Average range in temperature
(minimum to maximum)**

a What are the average minimum and maximum
temperatures in May?
maximum _____   minimum _____

b In January the average minimum temperature is
⁻10 °C.
The range in temperature is 11 °C.
Draw a bar on the graph to show this information.

3 Don't get burnt!

[Level 5]

A bottle of sunscreen has a pump on the top which squirts out
5 mℓ of sunscreen at a time.
The bottle contains 1 litre of sunscreen.
How many squirts of sunscreen are in the bottle? _____

Sun
screen

Paper 2 (cont.)

4 *Prime grid* *[SATs question Level 5]*

The diagram shows part of a number grid.
The grid has 6 columns.
All the **prime numbers** in the grid are circled.

a 35 is not circled.
Explain why 35 is **not** a prime number.

____

____

b The are no prime numbers circled in column Y.
Explain how you know there will **never** be a prime number in column Y.

____

____

____

c There is one prime number circled in column X.
Explain how you know there will **never** be another prime number in column X.

____

____

(43)	44	45	46	(47)	48
(37)	38	39	40	(41)	42
(31)	32	33	34	35	36
25	26	27	28	(29)	30
(19)	20	21	22	(23)	24
(13)	14	15	16	(17)	18
(7)	8	9	10	(11)	12
1	(2)	(3)	4	(5)	6
		↑ column X			↑ column Y

5 *How tall was he?* *[Level 5]*

Archaeologists are scientists who study the remains of people who lived in the past.
If they find a male's tibia (shin bone) they can use the rule below to work out the height of this male (in cm).

> **Multiply** the length of the tibia, in cm, by 2·5
> then **add** 76
> then **round** to the nearest whole number.

Two male skeletons are found.
The tibia from skeleton A is 43·8 cm long.
The tibia from skeleton B is 44·1 cm long.
An archaeologist says:
'The rule shows that the two males were the same height.'
Is she correct? Tick (✓) Yes or No.

☐ Yes ☐ No

Show working to explain your answer.

6 *Big winnings*

a Mrs O'Leary won £2625 in a lottery. *[Level 5]*
She spent 27% of her winnings on a new television.
How much is 27% of £2625? £________

b Mr Hamilton won £36 928. *[Level 6]*
He spent £1899 on a car.
What percentage of his winnings is that? ________%

Paper 2 (cont.)

7 *Same area* [Level 5]

a The square and the rectangle below have the **same area**.

4 cm

 2 cm — y cm —

Not drawn accurately

Work out the value of *y*. *y* = _____ cm

b The triangle and the rectangle below have the **same area**.

6 cm

4 cm

w cm

4 cm

Not drawn accurately

Work out the value of *w*.
Show your working.

w = _____ cm

8 *Straight lines* [SATs question Level 6]

The graph shows a straight line.

a Fill in the table for some of the points on the line.

(x, y)	(,)	(,)	(,)
x + y			

b Write an equation of the straight line. _____

c On the graph, draw the straight line that has the equation $x + y = 6$

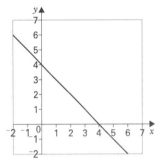

9 *Watch the bar!* [Level 6]

At a pony club fun day there are 15 horse jumps set up.
A successful jump earns the rider 3 points.
Knocking the bar down on a jump loses 1 point.
Any missed jumps score 0 points. A negative score is possible.

a What are the maximum and minimum points a rider could score?
maximum _____   minimum _____

b A rider attempts 10 jumps, 6 of these are successful.
How many points did she score? _____

c Complete this table to show 3 different ways to score 21 points.

Number of jumps attempted	7		
Number of successful jumps	7		
Number of bars knocked down	0		

Paper 2 (cont.)

10 Wind it up

[Level 6]

A crane has a long pipe on it. The end of the pipe is a circle.
The diameter of the circle is 8 cm.

8 cm

a What is the circumference of the circle? _____ cm

b 27 metres of rope is wound around the pipe.
About how many times does the rope go around the pipe?
Show your working and give your answer to the nearest
whole number.

8 cm

About _____ times around

11 Coughs and colds

The child's dose for a cough mixture can be worked out using this formula.

$$c = \frac{a}{a + 10} \times d$$

c = the child's dose, in mℓ
a = the age of the child, in years
d = the dose for an adult, in mℓ

a The adult dose for one cough mixture is 30 mℓ.
How much should a 5-year-old child take?
Show your working.

[Level 6]

_____

*b Jasmine is 10 years old. She takes 25 mℓ of another cough mixture.
How much of this cough mixture should her mother take?
Show your working.

[Level 7]

_____

12 Go Panthers!

[Level 7]

The Panthers rugby team have played 10 games so far this season.
The mean number of points they have scored is 25.
This list gives the number of points they scored in their first 9 games.
    26, 41, 16, 28, 15, 32, 21, 19, 34
How many points did they score in their 10th game? _____

13 Drink it up

[Level 7]

A fruit punch is made by mixing tea, ginger ale and juice in the ratio 2 : 4 : 9.

a If 6 litres of ginger ale are used, how much tea and juice should be used?

_____ ℓ __6__ ℓ _____ ℓ
   tea   ginger ale   juice

b Mary has 11 litres of juice. How much tea and ginger ale should she mix with it to make the punch?
Give your answer to 1 decimal place.

_____ ℓ _____ ℓ __11__ ℓ
   tea   ginger ale   juice

Answers – Number

1 Powers of ten   page 2

(A) $\frac{1}{10\,000}$, 0·0001, one ten thousandth

(B) **a** C   **b** A

(C) **a** **ii** 10^{-3} **iii** 10^{-9} **iv** 10^{-12}

b **ii** $5 \times \mathbf{10^9}$ bytes = $5 \times$ **1 000 000 000** bytes = **5 000 000 000** bytes

iii 8×10^{-3} seconds = $8 \times \frac{1}{1000}$ seconds = $\frac{8}{1000}$ seconds

or $8 \times \mathbf{10^{-3}}$ seconds = $8 \times \mathbf{0{\cdot}001}$ seconds = **0·008** seconds

c **i**

7 millimetres → 7×10^{-3} metres
7 centimetres → 7×10^{-2} metres
7 kilometres → 7×10^{3} metres
7 megametres → 7×10^{6} metres
7 nanometres → 7×10^{-9} metres

ii
5 milligrams → 5×10^{-3} grams
5 micrograms → 5×10^{-6} grams
5 kilograms → 5×10^{3} grams
＊50 kilograms → 5×10^{4} grams
＊500 nanograms → 5×10^{-7} grams

2 Multiplying and dividing by powers of ten   page 3

(A) NEIL ARMSTRONG STEPPED ON THE MOON WITH HIS LEFT FOOT FIRST.

(B) **a** £28   **b** 15·67 kg
c 37 500 cm^2 **d** 0·08 m^3

(C) $9{\cdot}75 \times 0{\cdot}01 = 9{\cdot}75 \times \frac{1}{100}$
$= 9{\cdot}75 \div 100$
So $9{\cdot}75 \times 0{\cdot}01$ is the same as $9{\cdot}75 \div 100$

3 Rounding   page 4

(A) **a**

Number	Nearest 100	Nearest 1000	Nearest 10 000
8 321 423	8 321 400	8 321 000	8 320 000
3 968 500	3 968 500	3 969 000	3 970 000

b

Number	Nearest whole number	to 1 d.p.	to 2 d.p.
12.473	12	12.5	12.47
19.987	20	20.0	19.99

(B) **a** *Possible answers are:*
i $900 \times 50 = 45\,000$ **ii** $400 \times 70 = 28\,000$
b 11·8 m^2 (1 d.p.) or 11·75 m^2 (2 d.p.)

(C) **a** Smallest 35 500, largest 36 499
b Any number from 35 650 to 35 749 is correct.

(D) **a** 14 800 000
b
Brazil	8 510 000
Argentina	2 770 000
Peru	1 290 000
Colombia	1 140 000
Bolivia	1 100 000
Total	14 810 000

c Because rounding then adding gives a less accurate estimate than adding then rounding.
d 17% (nearest per cent)
e About 99% (nearest per cent)

4 Adding, subtracting, multiplying and dividing integers   page 5

(A) **a**
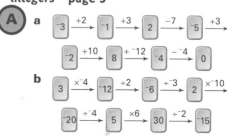

b

(B) **a** 10   **b** $^-7$ **c** 12   **d** $^-6$

(C) **a** $^-1 - {}^-4 = 3$ or $^-4 - {}^-7 = 3$ or $10 - 7 = 3$
b $^-4 \times 5 = {}^-20$, $10 \times {}^-2 = {}^-20$
c **i** $^-4 + {}^-1 + {}^-2 = {}^-7$
ii $^-4 + {}^-2 - 10 = {}^-16$

(D) **a** $^-4, 6$ **b** $^-2, {}^-7$

5 Prime factor decomposition   page 6

(A) **a**
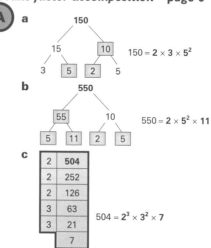

$150 = 2 \times 3 \times 5^2$

b

$550 = 2 \times 5^2 \times 11$

c
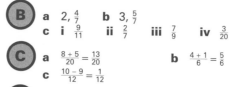

2	504
2	252
2	126
3	63
3	21
	7

$504 = 2^3 \times 3^2 \times 7$

(B) **a** $2, \frac{4}{7}$ **b** $3, \frac{5}{7}$
c **i** $\frac{9}{11}$ **ii** $\frac{2}{7}$ **iii** $\frac{7}{9}$ **iv** $\frac{3}{20}$

(C) **a** $\frac{8+5}{20} = \frac{13}{20}$ **b** $\frac{4+1}{6} = \frac{5}{6}$
c $\frac{10-9}{12} = \frac{1}{12}$

(D) **a** *Possible answers are:* 4, 12, 20, 28, 36, ...
b 72, 56

Number

6 Common factors of algebraic expressions page 7

A
a 3x and 7x
8x and 4x
4x² and x²
2x² and 8x
3xy and 6xz

matches to: x², 4x, x, 3x, 2x, 2ab, b, 3c, 2a, a

b 4ab and 5ac
8ac and 6ab
5a²b and bc
6a²c and 3bc
4ab² and 6a²b

B
a y   **b** s   **c** x   **d** 5   **e** 3m
f 3r   **g** ab   **h** yz   **i** 3bc

C

5x	6x	8	3ab	6a²
x	4n	4	3a	5a²c
3x²	21n	14n²	7n	10c²
9x	3x	7	6cn	5c
8xy	xy	21x²y	2c	8c²

7 is left

D
a Many possible answers. Three are 3x, 9x, 6xy.
b Many possible answers. Three are 2qr, 6q²r, 10qrt.
c Many possible answers. Two are
  ● 10a² and 15a²b
  ● 20a³ and 5a²c².

7 Divisibility and algebra page 8

A
a Possible answers are: 18, 36, 45, 54, 72 and 90
b $10t + u = 10t + (9 - t)$
$= 10t + 9 - t$
$= 9t + 9$
$= 9(t + 1)$
$9(t + 1)$ is always divisible by 9.

B
a The number is $10t + 0$.
So the number is $10t$, which is always divisible by 10.
b The number is $10t + 5 = 5(2t + 1)$
$5(2t + 1)$ is always divisible by 5.
c The number is $100h + 10t + 5 = 5(20h + 2t + 1)$
$5(20h + 2t + 1)$ is always divisible by 5.

8 Powers page 9

A
a 4096   **b** 7·29   **c** 4096
d 2541·17 (2 d.p.)   **e** ⁻343   **f** 47·05 (2 d.p.)
g $\frac{1}{125}$ or 0·008   **h** 40 000

B
a **i** 1   **ii** 1   **iii** 1
b **i** 2   **ii** 1   **iii** ⁻3   **iv** ⁻1   **v** 0
c 2⁶, 4³
d 4
e $5^2 \times 2^4 = 400$
f $1^3 + 4^3 = 65$

C
a F   **b** F   **c** F   **d** F
e T   **f** F   **g** T

D
a 4⁴   **b** 4⁴   **c** 3⁴ and 3⁶

9 Multiplying and dividing numbers with indices page 10

A
a $(3 \times 3) \times (3 \times 3 \times 3 \times 3) = 3^6$
b $(7 \times 7 \times 7) \times (7 \times 7 \times 7 \times 7 \times 7) = 7^8$
c $\frac{4 \times 4 \times 4 \times 4 \times 4}{4 \times 4 \times 4} = 4^2$
d $\frac{6 \times 6 \times 6 \times 6 \times 6 \times 6}{6 \times 6 \times 6 \times 6 \times 6} = 6^1$ or 6

B
b 4⁷   **c** 4¹⁵   **d** 4³   **e** 6⁶   **f** 6⁸
g 6²   **h** 6¹⁶   **i** 5¹⁰   **j** 5⁰   **k** 5⁹
l 5¹¹   **m** 4⁸   **n** 5⁵   **o** 6¹²   **p** 6⁹
Due to Earth's gravity it is impossible for mountains to be higher than 15 km.

C
a 2⁶   **b** 4⁴   **c** $7^0 \times 7^3$ **d** $\frac{5^9}{5^3 \times 5^3}$

D No, because the base numbers are not the same so you cannot add the indices.

E
a a⁷   **b** y¹⁰   **c** p²   **d** r⁵
e x³   **f** z⁷   **g** b⁹   **h** d¹¹

10 Square roots and cube roots page 11

A
a **i** 3   **ii** 5 or ⁻5   **iii** 8 or ⁻8
iv 11   **v** 2
b **i** Positive   **ii** Negative
c **i** $\sqrt{225} = \sqrt{9 \times 25}$
$= \sqrt{9} \times \sqrt{25}$
$= 3 \times 5$
$= 15$
ii $\sqrt{324} = \sqrt{4 \times 81}$ or $\sqrt{324} = \sqrt{9 \times 36}$
$= \sqrt{4} \times \sqrt{81}$ $= \sqrt{9} \times \sqrt{36}$
$= 2 \times 9$ $= 3 \times 6$
$= 18$ $= 18$

B
a 21·86   **b** 2·35   **c** 7·68
d 6·54   **e** 10·53

C $\sqrt{4} \times \sqrt{25} = \sqrt{100}$ $\sqrt{9} \times \sqrt{9} = \sqrt{81}$

D Possible answer is
$\sqrt{40}$ lies between 6 and 7.
Try 6·5
$6·5 \times 6·5 = 42·25$ too big
$6·4 \times 6·4 = 40·96$ too big
$6·3 \times 6·3 = 39·69$ too small
$6·32 \times 6·32 = 39·94$ too small
$6·33 \times 6·33 = 40·07$ too big
$6·325 \times 6·325 = 40·006$.
$\sqrt{40}$ is between 6·32 and 6·325.

11 Multiplying and dividing by numbers between 0 and 1 page 12

A $50 \times 0·3$, $50 \times 0·9$, $50 \div 28$, $50 \div 1·6$, $50 \times 0·04$, $50 \div 2\frac{1}{4}$

B

START	0·08 ÷ ½	0·08 ÷ 0·1	0·08 ÷ 0·5	0·08 × 0·09	0·08 × 1·9	0·08 × 8/9	
	0·08 × ½	0·08 × 10	0·08 ÷ 5	0·08 ÷ 0·09	0·08 ÷ 1·9	0·08 × 8/9	
	0·08 × 2	0·08 × 0·1	0·08 ÷ 0·05	0·08 × 90·9	0·08 × 0·9	0·08 × 8·9	FINISH

C
a **i** 1800, 180, 18, 1·8, 0·18
ii 4000, 40 000, 400 000
b **i** 0·018   **ii** 40 000 000

D
a **i** Greater   **ii** Less
b **i** ÷   **ii** ×   **iii** ×

12 Order of operations   page 13

(A)
1 E. 46 E. 6 U. 7 L. 38 R. 51   EULER
2 A. 52 L. 55 S. 27 P. 4 A. 10 C. 40   PASCAL
3 E. 7 V. 4 I. 5 T. 12 A. 64   VIETA

(B)
a A is the square of the sum of 3 and 5, and B is the square of 5 added to 3.
b A is the square root of the sum of 16 and 9, and 8 is the square root of 16 plus 9.

(C)
a $7 \times (2 + 3) = 35$
b $(4 + 3)^2 - 10 \times 3 = 19$
c $(9 + 6) \div (7 - 2) = 3$

(D)
a $76 = (2 + 7)^2 - 5$ b $(1 + 5)^2 - (3 \times 7) = 15$

13 Mental calculations   page 14

(A)
a 31   b $^-5$ c $^-11$ d 14
e 0·75   f $\frac{3}{5}$ g 40%   h 114·5%
i £55   j 255 mm

(B)
IN THE MIDDLE AGES SUGAR COST NINE TIMES AS MUCH AS MILK

(C)
a £180   b £43·40   c £16·25

(D)
a i 11·4   ii 19
b $50 \times 43 = 2150$

14 Solving word problems mentally   page 15

(A)
a 2   b 54   c 125 cm³   d 59°
e 5·8 kg   f 86 cm²   g 90 km   h 14 cm
i 40%   j 175

(B)
a 60p   b 3   c 6   d 288   e £1050

15 Estimating   page 16

(A)
a B   b C   c A   d B   e A

(B)
a Karen   b Joy   c Gulam

(C)
Possible answers are:
a $50 \times 50 = 2500$ b $6 \times 11 = 66$
c $64 \div 8 = 8$ or $63 \div 7 = 9$ d $\frac{8 + 4}{4} = 3$

(D)
a Estimate could be $10 \times 50 = 500$ m² or $12 \times 50 = 600$ m²
   Accurate answer = 593·92 m²
b Estimate could be $\frac{300}{15} = 20$ pounds
   Accurate answer = 19·3 pounds (1 d.p.)

16 Written calculation   page 17

(A)
a (5751·84)   5741·84   5741·84
b 1288   1288   (1287·06)
c (5·355)   5·4   5·4

(B)
a i C   ii B   iii D
b Because the last digit must be 4 and $317 \times 2·2$ will be greater than $300 \times 2 = 600$.

(C)
a 2376·7   b 278·6
c 1356·7   d 18·6

(D)
a 1·71 m   b 18·093 m²
c 4·7 m   d 13 m

17 Choosing a strategy for calculation   page 18

(A)
a £28·25   b 18p or £0·18   c £42

(B)
100

(C)
20 m³ truck, £5 cheaper

(D)
a 2 mins 45 secs   b 1 min 53 secs

(E)
27p

18 Using a calculator   page 19

(A)
a $\boxed{x^2}$ should be $\boxed{\wedge}$ or $\boxed{x^y}$.
b Shaun should have keyed $\boxed{a^{b/c}}$ between the 4 and the 2.
c Shaun needs an extra $\boxed{)}$ before the $\boxed{\times}$ and the $\boxed{=}$.
d $\boxed{)}$ is in the wrong place. It should be after $\boxed{\pi}$.

(B)

(C)
a $6\frac{14}{15}$ b $1\frac{15}{28}$ c $28\frac{8}{9}$
d $1\frac{19}{21}$ e $^-9\frac{5}{14}$ f $^-39\frac{23}{63}$

19 Fractions, decimals and percentages   page 20

(A)
a $\frac{4}{5}$ b $\frac{3}{4}$ c $\frac{2}{3}$ d $1\frac{1}{3}$ e $\frac{1}{10}$
f $\frac{1}{50}$ g $\frac{1}{25}$ h $\frac{2}{3}$ i $\frac{5}{9}$ j $\frac{1}{1}$ or 1

(B)
a i 20%   ii 42·5%
b 0·76
c $\frac{4}{7}$

(C)
a i $\frac{1}{4}$ ii $\frac{11}{28}$
b 64% (nearest percent)

(D)
a 50%   b 40%
c 37·5%   d 62·5%

20 Comparing proportions   page 21

(A)
a <   b >   c <   d <   e <
f $\frac{1}{2}$, $\frac{5}{8}$, $\frac{2}{3}$, $\frac{3}{4}$ g 55%, $\frac{3}{5}$, $\frac{13}{20}$, 0·68
h 0·3, $\frac{1}{3}$, 44%, $\frac{4}{9}$ i $\frac{7}{25}$, 29%, $\frac{9}{30}$, 0·31

(B)
a Ross
b Gareth
c No, because $\frac{4}{10}$ is not the same proportion as $\frac{4}{200}$.
d Manzoor

(C)
a 76·1%
b $\frac{186}{257}$ = 72·4% of the outside seeds germinated. This is less than 76·1% for the inside seeds.

21 Adding and subtracting fractions   page 22

(A)
a **i**

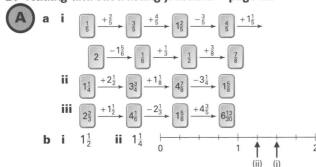

ii $1\frac{1}{4} \xrightarrow{+2\frac{1}{2}} 3\frac{3}{4} \xrightarrow{+1\frac{1}{8}} 4\frac{7}{8} \xrightarrow{-3\frac{1}{4}} 1\frac{5}{8}$

iii $2\frac{2}{3} \xrightarrow{+1\frac{1}{2}} 4\frac{1}{6} \xrightarrow{-2\frac{1}{3}} 1\frac{5}{6} \xrightarrow{+4\frac{3}{5}} 6\frac{13}{30}$

b **i** $1\frac{1}{2}$ **ii** $1\frac{1}{4}$ (number line 0 to 2, arrows at (ii) and (i) between 1 and 2)

c **i** $\frac{79}{105}$ **ii** $\frac{283}{495}$ **iii** $1\frac{63}{65}$ **iv** $3\frac{55}{84}$

(B)
a $\frac{1}{3}$ kg   **b** $\frac{13}{15}$ m

(C)
a **i** 3   **ii** 6   **iii** 6, 7
b **i** $3\frac{1}{2}$ **ii** 6
c **i**

$\frac{1}{4}$	$\frac{5}{8}$	$\frac{5}{8}$
$\frac{7}{8}$	$\frac{1}{2}$	$\frac{1}{8}$
$\frac{3}{8}$	$\frac{3}{8}$	$\frac{3}{4}$

ii

$1\frac{1}{6}$	$2\frac{5}{6}$	$3\frac{1}{2}$
$4\frac{5}{6}$	$2\frac{1}{2}$	$\frac{1}{6}$
$1\frac{1}{2}$	$2\frac{5}{6}$	$3\frac{5}{6}$

iii

$1\frac{1}{10}$	$3\frac{3}{5}$	$3\frac{3}{5}$
$4\frac{3}{5}$	$2\frac{1}{2}$	$\frac{2}{5}$
$1\frac{4}{5}$	$1\frac{4}{5}$	$3\frac{9}{10}$

22 Fraction and percentage of   page 23

(A) THE SAHARA DESERT EXPANDS BY ABOUT ONE KM PER MONTH

(B)
a **i** $\frac{3}{5}$ of 35   **ii** $\frac{4}{9}$ of 81   **iii** 30% of 210
b **i** 84% of 56   **ii** $17\frac{1}{2}$% of £13
iii 58·3% of 2·1 kg

(C)
a 80   **b** 120   **c** 45

(D)
a 40   **b** £60

23 Multiplying fractions   page 24

(A)
a (tree: $\frac{9}{80}$; $\frac{3}{8}$, $\frac{3}{10}$; $\frac{3}{4}$, $\frac{1}{2}$, $\frac{3}{5}$)
b (tree: $\frac{1}{5}$; $\frac{3}{5}$, $\frac{1}{3}$; $\frac{3}{4}$, $\frac{4}{5}$, $\frac{5}{12}$)
c (tree: $1\frac{1}{14}$; $\frac{5}{6}$, $1\frac{2}{7}$; $\frac{20}{9}$, $\frac{3}{8}$, $\frac{24}{7}$)

(B)
a 4 m^2 **b** 8 m^2 **c** $4\frac{1}{8}$ m^2 **d** $3\frac{13}{36}$ m^2

(C)
a T   **b** F   **c** T   **d** F   **e** T
f F   **g** F   **h** T   **i** T

(D)
a **i** Many possible answers. Three are $\frac{3}{4} \times \frac{3}{4}$, $\frac{3}{8} \times \frac{3}{2}$, $\frac{9}{8} \times \frac{1}{2}$.
ii Many possible answers. Three are $\frac{7}{5} \times \frac{2}{3}$, $\frac{7}{3} \times \frac{2}{5}$, $\frac{14}{3} \times \frac{1}{5}$.
b **i** $\frac{3}{4} \times \frac{4}{5} \times \frac{5}{6}$ **ii** $\frac{4}{5} \times \frac{6}{7} \times \frac{7}{8}$

24 Dividing fractions   page 25

(A)
a $\frac{1}{5}$ **b** 18   **c** $\frac{1}{14}$ **d** 25
e $6\frac{2}{3}$ **f** 27   **g** 2   **h** $1\frac{1}{9}$
i $\frac{12}{25}$ **j** $6\frac{1}{4}$ **k** $6\frac{2}{5}$ **l** $4\frac{1}{8}$
m $4\frac{7}{12}$ **n** $\frac{3}{4}$ **o** $\frac{16}{35}$ **p** $4\frac{1}{2}$
Summer on Uranus lasts for 21 years.

(B)
a 2   **b** $3\frac{5}{7}$ **c** $\frac{3}{10}$

(C)
a 52   **b** 8   **c** £$3\frac{1}{5}$ or £3·20

(D)
a Many possible answers. Three are $\frac{1}{2} \div \frac{4}{3}$, $\frac{1}{4} \div \frac{2}{3}$, $\frac{3}{16} \div \frac{1}{2}$.
b Three of $\frac{4}{3} \div \frac{6}{4}$, $\frac{4}{6} \div \frac{3}{4}$, $\frac{4}{9} \div \frac{2}{4}$, $\frac{4}{2} \div \frac{9}{4}$, $\frac{4}{18} \div \frac{1}{4}$, $\frac{4}{1} \div \frac{18}{4}$.

25 Percentage change   page 26

(A)
a **i** £4·20   **ii** £3·15
b 26·5% (1 d.p.)   **c** 320 g

(B)
a 162
b **i** 11 309
ii 25·6% (1 d.p.), 67·0% (1 d.p.)
c 584

(C)
a D
b Calculation A: What is 45 increased by 70%?
Calculation B: What is 7% of 45?
Calculation C: What is 70% of 45?

26 Proportionality   page 27

(A)
a £21   **b** £30   **c** 27·5 minutes
d £68·80

(B)

Tuna pasta (for 10 people)	
• 125 g butter	• 1250 g pasta
• 5 tbsp flour	• 475 g tuna
• 950 mℓ milk	• 625 g grated cheese

(C)
a 6 g   **b** Spaghetti, 0·84 g

(D)
a 0·949 Euros   **b** £5   **c** 7·19 Danish Krona

27 Ratio   page 28

(A)
a 2 : 3   **b** 6 : 7   **c** 3 : 6 : 7   **d** 5 : 14
e 8 : 5   **f** 1 : 4   **g** 1 : 4   **h** 3 : 2
i 1 : 2·75   **j** 1 : 12·6   **k** 5·43   **l** 1 : 1·24

(B)
a 7 : 5   **b** 5 : 4

(C)
a 1 : 3
b **i** A 2·4 : 1   B 2·25 : 1   **ii** A
c Heathcote United, because 1 female for every 5·38 males is greater than 1 female for every 5·47 males.

28 Ratio and proportion problems page 29

 a **i** 1 : 2   **ii** 9 : 2
     b **i** $\frac{1}{10}$, 10%, 0·1   **ii** $\frac{9}{20}$, 45%, 0·45

 a 300 mℓ   **b** 90 mℓ
     c No, it is 1 part in 6, or $\frac{1}{6}$, or 16·67%, not 20%.

C **a** 400 m^2 **b** 45°, 75°, 60°   **c** 38
     d 72   **e** 420 cm
     f 4 : 5

Answers – Algebra

29 Starting algebra   page 31

A

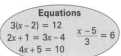

Equations
$3(x - 2) = 12$
$2x + 1 = 3x - 4$ $\frac{x-5}{3} = 6$
$4x + 5 = 10$

Formulae
$P = I^2R$ $a = \frac{d}{t^2}$
$V = \frac{1}{3}Ah$

Functions
$y = 3x + 2$ $y = \frac{x+6}{4}$
$y = 5x - 1$ $y = 20 - \frac{x}{3}$

B **a i** $6(1 + 2) = 18$
$6 \times 1 + 12 = 18$
ii $6(8 + 2) = 60$
$6 \times 8 + 12 = 60$
iii $6(^-2 + 2) = 0$
$6 \times ^-2 + 12 = 0$
b i YES   **ii** YES

C **a** False, because it doesn't give the relationship between two variables.
b True, because each variable stands for something specific.
c False, because the left hand side does not equal the right hand side for any value of x.

30 Inequalities   page 32

A **a** $c > 7$ **b** $t \geqslant 25$ **c** $p \leqslant 28$
d $g \leqslant 9$ **e** $0.5 \leqslant h < 3$

B **a** 1, 3, 5, 7, 9   **b** 2, 4, 6
c 9, 16, 25, 36, 49   **d** $^-4, ^-3, ^-2, ^-1$

C **a** Yes
b i Yes   **ii** Yes   **iii** Yes
iv Yes   **v** No   **vi** No
c Reverse the inequality sign.

31 Solving equations   page 33

A SOUND TRAVELS FASTER THROUGH WATER THAN AIR.

B **a** $14 + n + n + 17 = 53$, $n = 11$
b $4(p + 7) = 64$, $p = 9$

C **a i** 34   **ii** 3
b $2g + 3h + 4p - 2q = 23$

32 Equations with unknowns on both sides   page 34

A **a** $n = 2$ **b** $n = 4$ **c** $n = 2\frac{1}{2}$
d $n = 7$ **e** $n = 3\frac{1}{3}$ **f** $n = \frac{4}{5}$
g $n = 1\frac{1}{2}$ **h** $n = 3$ **i** $n = 8$
j $n = ^-1$ **k** $n = \frac{1}{4}$
The top speed of the first train was 5 miles per hour.

B **a** $6x - 2 = 5x + 4$, $x = 6$
b $4x + 3 = 7x - 6$, $x = 3$, Area = 90 cm^2

C $3n - 8 = 20 - n$, number = 7

D **a** False   **b** False   **c** True
d True   **e** True   **f** True

33 Solving non-linear equations   page 35

A **a** $y = +9$ or $^-9$ **b** $a = +12$ or $^-12$
c $c = +3$ or $^-3$

B **a** $m = +6.2$ or $^-6.2$ **b** $x = 13$
c $z = 0.8$

C **a i**

Try	$x^2 - x$	Comment
$x = 8.25$	59.8125	close but too small
$x = 8.27$	60.1229	close but too big
$x = 8.26$	59.9676	very close but too small

ii $x = 8.26$
b $y = 3.32$

34 Algebra and proportion   page 36

A **a**

Glue 1 (mℓ)	2	4	6	8	10
Glue 2 (mℓ)	3	6	9	12	15

b Yes, because the ratio $\frac{\text{glue 1}}{\text{glue 2}}$ is constant for all values.

c
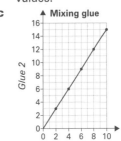
Mixing glue

d Yes, because the ratio $\frac{\text{glue 1}}{\text{glue 2}}$ is constant.
e $s = \frac{3}{2}f$

B **a**
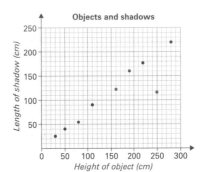
Objects and shadows

b Because measurement data is never exact, and it is possible to get errors in measurement.
c Yes, because all of the points except (250, 116) lie in an almost straight line.

C $\frac{4}{7} = \frac{x}{280}$, $x = 160$ mℓ

35 Collecting like terms   page 37

A   a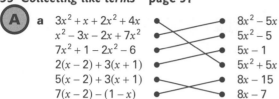

$3x^2 + x + 2x^2 + 4x$
$x^2 - 3x - 2x + 7x^2$
$7x^2 + 1 - 2x^2 - 6$
$2(x - 2) + 3(x + 1)$
$5(x - 2) + 3(x + 1)$
$7(x - 2) - (1 - x)$

$8x^2 - 5x$
$5x^2 - 5$
$5x - 1$
$5x^2 + 5x$
$8x - 15$
$8x - 7$

   b   **i** $3e^2 - 5 - 2e^2$ **ii** $6c + 4 - 2c - 1$

B   a $p = 10n + 1$ **b** $p = 16m - 6$
   c   Length 1 = $6p + 4$, length 2 = $2p + 6$

C   a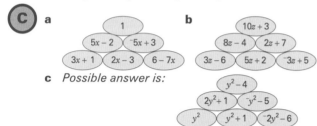

```
        1
    5x - 2   5x + 3
 3x + 1  2x - 3  6 - 7x
```

   b
```
        10z + 3
    8z - 4   2z + 7
 3z - 6  5z + 2  ⁻3z + 5
```

   c   Possible answer is:
```
        y² - 4
    2y² + 1   ⁻y² - 5
  y²   y² + 1   ⁻2y² - 6
```

36 Multiplying and dividing expressions   page 38

A
   a $15m$ b $24m$ c ^-8m d ^-12m
   e $6m^2$ f $32m^2$ g $12mn$ h m
   i $\frac{7m}{5}$ j m^4 k m^2 l m^7
   m $4m$ n $4m^2$ o $6m^6$ p $10m^3$
   q $30m^{10}$ r $\frac{4m^2}{3}$
   Your brain is 75% water.

B   a $15n \times n$ **b** $4b \times 3b^2$ **c** $\frac{15p^6}{10p^2}$

C   a **i** $A = 15x^2$ **ii** $A = 9x$ **iii** $A = 32x^2$
   b **i** $l = 6$ **ii** $l = 3x$

37 Writing expressions   page 39

A   a $5x - 1$ **b** $4x - 1$ **c** 58

B   a
```
   n
   ↓
  n + 6
   ↓
 2n + 6
   ↓
 n + 3
   ↓
   n
```
   b
```
   n
   ↓
  3n
   ↓
 3n + 8
   ↓
 4n + 8
   ↓
 n + 2
   ↓
   n
```

C   a $A = 21ab$, $P = 14a + 6b$
   b $9a$ by $2a$
   c **i** Perimeter of $A = 8x$
      Perimeter of $B = 6x + 4$
   ii $x = 2$

38 Factorising   page 40

A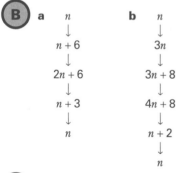

3x + 27	3(x + 9)	5(x + 2)	2(x − 6)	2x − 8
3(x + 4)	3x + 12	5x + 10	2x − 12	2(x − 4)
3x + 9	5x + 25	5(x + 5)	2(x − 2)	2x − 4
3(x + 3)	3x + 6	3(x + 2)	2x − 2	2(x − 1)

B   a $4(x + 2)$ **b** $3(n - 4)$ **c** $5(y + 1)$
   d $5(5a - 3)$ **e** $4(4x + 3)$ **f** $8(3 + n)$
   g $2(2 + 5x)$ **h** $6(3n - 2)$ **i** $3(x + y)$
   j $6(a + 2b)$ **k** $y(3y + 1)$ **l** $x(x - 5)$

C   a $2(x + 3)$ **b** $6(2y - 1)$ **c** $5(3a + 2)$
   d $7(2 + 3n)$ **e** $5(7a - 5)$ **f** $3(7x + 6)$
   g $6(6 - 5a)$ **h** $x(x + 3)$ **i** $a^2(a + 1)$
   j $5y(y + 1)$ **k** $3n(2n + 1)$ **l** $a^2(4a - 1)$

D

Expression	a $8x + 4$	b $6a - 12$	c $48x + 24$
Annie	~~8(x + 1)~~	~~3(a - 4)~~	~~12(x + 2)~~
Bob	$2(4x + 2)$	$6(a - 2)$	$6(8x + 4)$
Carla	$4(2x + 1)$	$3(2a - 4)$	$24(2x + 1)$

Expression	d $45 - 15n$	*e $y^3 + y^2$	*f $6n^3 - n^2$
Annie	$5(9 - 3n)$	~~y³(1 + y)~~	$n^2(6n - 1)$
Bob	$15(3 - n)$	$y(y^2 + y)$	~~6n²(n − 1)~~
Carla	~~5(9 − 5n)~~	$y^2(y + 1)$	$n(6n^2 - n)$

39 Adding and subtracting algebraic fractions   page 41

A   a $\frac{7}{9}$ b $\frac{7}{x}$ c $\frac{4}{y}$ d $\frac{3y}{5}$
   e $\frac{3}{6} + \frac{2}{6} = \frac{5}{6}$ f $\frac{16}{24} - \frac{15}{24} = \frac{1}{24}$
   g $\frac{3b}{ab} + \frac{4a}{ab} = \frac{3b + 4a}{ab}$ h $\frac{4c}{12} + \frac{3d}{12} = \frac{4c + 3d}{12}$
   i $\frac{sp}{qs} + \frac{rq}{qs} = \frac{sp + rq}{qs}$

B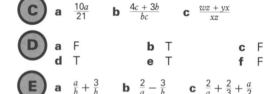

```
 x   y        2   5        x   y            2   y        2   y
--- + ---    --- + ---    --- − ---        --- − ---    --- + ---
 2   5        x   y        2   5            x   5        x   5

5x − 2y      5x + 2y      2y + 5x          10 + xy      10 − xy
-------      -------      -------          -------      -------
  10           10           xy               5x           5x
```

C   a $\frac{10a}{21}$ b $\frac{4c + 3b}{bc}$ c $\frac{wz + yx}{xz}$

D   a F   b T   c F
   d T   e T   f F

E   a $\frac{a}{b} + \frac{3}{b}$ b $\frac{2}{a} - \frac{3}{b}$ c $\frac{2}{a} + \frac{2}{3} + \frac{a}{2}$

40 Substituting into expressions   page 42

A   a **i** 15   **ii** 7   **iii** 44   **iv** 27   **v** 20
   b **i** ⁻8   **ii** ⁻7   **iii** 3   **iv** 10   **v** 75

B   a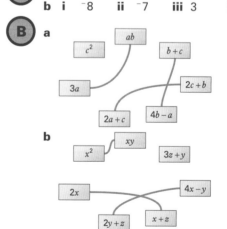

```
           ab
   c²              b + c

  3a              2c + b

     2a + c   4b − a
```

   b
```
        xy
   x²          3z + y

   2x              4x − y

      2y + z   x + z
```

C   a 28   b 130   c 990

Must be odd	Must be even	Can't tell
$3y$ y^2	$2y$	$\dfrac{y-1}{2}$
$(y+2)$ $(y-2)$	$5y+1$	

41 Substituting into formulae page 43

 a $p = £11$ **b** £20 **c** £29·50

B **a** $A = 340$ cm^2 **b** $A = 0·6$ m^2
c $h = 0·6$ m

C **a** $p = 108$ watts **b** $R = 2·5$ ohms

D **a** **i** $w = 18·29$ g (2 d.p.)
  ii 40·17 g (2 d.p.)
  iii 135·59 g (2 d.p.)
b 5·19 cm (2 d.p.)

E **a** $h = 5$ cm
b $h = 9·22$ cm (2 d.p.)
c $h = 5·66$ cm (2 d.p.)
d $h = 0·33$ m (2 d.p.)

42 Changing the subject of a formula page 44

A **a** $d \rightarrow \boxed{\text{divide by } t} \rightarrow \dfrac{d}{t}$
  $st \leftarrow \boxed{\text{multiply by } t} \leftarrow s$
  so $d = st$
b **i** $d = 360$ km **ii** $d = 79·5$ km

B **a** $h \rightarrow \boxed{\text{multiply by } A} \rightarrow \boxed{\text{divide by } 3} \rightarrow \dfrac{Ah}{3}$
  $\dfrac{3V}{A} \leftarrow \boxed{\text{divide by } A} \leftarrow \boxed{\text{multiply by } 3} \leftarrow V$
  so $h = \dfrac{3V}{A}$
b **i** $h = 8$ mm **ii** $h = 12$ cm

C **a** $D = \dfrac{C}{\pi}$
b $l = d - 3t$ $t = \dfrac{d-l}{3}$
c $P = \dfrac{100I}{RT}$ $R = \dfrac{100I}{PT}$ $T = \dfrac{100I}{PR}$

D
$d = 2ef$	$d = \dfrac{e+f}{2}$	$d = 2(e+f)$	$d = \dfrac{ef}{2}$
$e = 2d - f$	$e = \dfrac{2d}{f}$	$e = \dfrac{d}{2f}$	$e = \dfrac{d}{2} - f$

E **a** $r = \dfrac{t}{2} - s$ **b** $f = \sqrt{e+g}$

43 Finding formulae page 45

A Shaded area
  = large rectangle area – small rectangle area
  = $3a \times 2b - a \times b$
  = $6ab - ab$
  = $5ab$

B **a**

Shape		Number of faces (f)	Number of vertices (v)	$f+v$	Number of edges (e)
Cube		6	8	14	12
Triangle-based pyramid		4	4	8	6
Cone		2	1	3	1
Triangular prism		5	6	11	9
Hexagonal prism		8	12	20	18

b $e = f + v - 2$
c **i** Yes **ii** No

44 Generating sequences page 46

A **a** **i** 3, 6, 12, 24, 48, 96 **ii** 7, 4, 1, ⁻2, ⁻5, ⁻8
  iii 2, 2·4, 2·8, 3·2, 3·6, 4·0 **iv** 5, 6, 8, 11, 15, 20
  v 2, 3, 5, 8, 13, 21
b **ii–iii** The difference between consecutive
    terms is constant.

B

Rule	$T(1)$	$T(2)$	$T(3)$	$T(4)$	$T(5)$	...	$T(20)$
$T(n) = 30 - 2n$	28	26	24	22	20	...	⁻10
$T(n) = 3n + 0·5$	3·5	6·5	9·5	12·5	15·5	...	60·5
$T(n) = \dfrac{n}{10}$	0·1	0·2	0·3	0·4	0·5	...	2

C **a** 1, 3, 2, ⁻1, ⁻3, ⁻2, 1, 3, 2, ⁻1, ⁻3, ⁻2
b The pattern 1, 3, 2, ⁻1, ⁻3, ⁻2 repeats.

45 Describing and continuing sequences page 47

A **a** **i** 1·7, 2·1, 2·5; rule: add 0·4
  ii 40 000, ⁻400 000, 4 000 000; rule:
    multiply by ⁻10
  iii 6, 3, 1·5; rule: divide by 2
  iv 34, 55, 89; rule: add previous two terms
b Possible answers are 2, 4, 6, 8, 10, 12; rule: start
  at 2 and add 2 to the previous term, and
  2, 4, 6, 10, 16, 26; rule: begin with the first terms
  2, 4 and add the previous 2 terms together.

B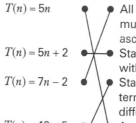
$T(n) = 5n$ — All terms two less than a multiple of 7. Starts at 5 and ascends.
$T(n) = 5n + 2$ — Starts at 7. Ascending terms with common difference of 5.
$T(n) = 7n - 2$ — Starts at 35. Descending terms with common difference of 7.
$T(n) = 40 - 5n$ — Ascending multiples of 5.
$T(n) = 42 - 7n$ — Multiples of 5. Starts at 25 and descends.

C **b** Multiples of 6, ascending.
c Each term is one more than a multiple of 6. It
  starts at 7 and ascends.
d Multiples of 6, starting at 60 and descending.

D Many possible answers. Some are:
First term: 4, rule: add 4
or First term: 8, rule: multiply by 4

46 Quadratic sequences   page 48

A

Rule	T(1)	T(2)	T(3)	T(4)	T(5)	T(6)
$T(n) = n^2$	1	4	9	16	25	36
$T(n) = n^2 + 2$	3	6	11	18	27	38
$T(n) = 2n^2 + 4$	6	12	22	36	54	76
$T(n) = n^2 + n$	2	6	12	20	30	42
$T(n) = 10 - 2n^2$	8	2	⁻8	⁻22	⁻40	⁻62
$T(n) = n^2 + 4n - 5$	0	7	16	27	40	55

B   **a** **i**   6,   9,   14,   21,   30,   …

   3   5   7   9

   2   2   2

ii   4,   6,   9,   14,   22,   …

   2   3   5   8

   1   2   3

iii   3,   7,   13,   21,   31,   …

   4   6   8   10

   2   2   2

b **i** and **iii**, because the second differences are all the same.

C   **a** 407   **b** ⁻98

D   **a** $\frac{1}{4}, \frac{2}{7}, \frac{1}{4}, \frac{4}{19}$ **b** $\frac{1}{2}, \frac{3}{5}, \frac{1}{2}, \frac{7}{17}$

47 Sequences in practical situations   page 49

A   **a** **i**   The white tile
      ii   The purple tiles
   b 1 white and 45 purple
   c 50

B   **a** 6 hexagonal shapes are added, one to each arm
   b $H = 6n + 1$

C   *Possible answer is:*

D   **a**

Level	1	2	3	4	5
Number of blocks	7	12	17	22	27

   b First term: 7, rule: add 5
   c Number of blocks = $5n + 2$
   d **i** 52   **ii** 102

48 Finding the rule for the *n*th term   page 50

A   **a**

Term	6	8	10	12	14
Difference		2	2	2	2

   $T(n) = \underline{2n + 4}$ $T(n) = \underline{104}$

   b

Term	27	32	37	42	47
Difference		5	5	5	5

   $T(n) = \underline{5n + 22}$ $T(50) = \underline{272}$

   c

Term	4	3	2	1	0
Difference		⁻1	⁻1	⁻1	⁻1

   $T(n) = \underline{⁻n + 5}$ $T(50) = \underline{⁻45}$
   or $T(n) = 5 - n$

d

Term	2·1	2·2	2·3	2·4	2·5
Difference		0·1	0·1	0·1	0·1

   $T(n) = \underline{0·1n + 2}$ $T(50) = \underline{7}$

B   **a** $T(n) = 5n + 1$, 20
   b $T(n) = ⁻2n + 10$ or $T(n) = 10 - 2n$, 50

C   **a**

Number of squares	1	2	3	4	5
Number of matches	4	7	10	13	16

   b 76
   c 50
   d No, because the rule is $T(n) = 3n + 1$, and there is no whole number *n* that makes $201 = 3n + 1$ true.

49 Functions   page 51

A   **a**

Input	5	12	3	10
Output	13	34	7	28

   b

Input	8	200	12	20
Output	9	57	10	12

B   **a**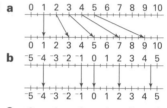
   b
   c 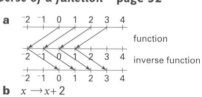   Identity, ⁻41·9

C   **a** add 5   **b** multiply by 3

D   **a** C   **b** E   **c** D

50 Inverse of a function   page 52

A   **a**
   function
   inverse function

   b $x \longrightarrow x + 2$

B   **a** $\frac{x}{3} \longleftarrow$ divide by 3 $\longleftarrow x$
      The inverse of $x \longrightarrow 3x$ is $x \longrightarrow \frac{x}{3}$.
   b $4(x + 3) \longleftarrow$ multiply by 4 $\longleftarrow$ add 3 $\longleftarrow x$
      The inverse of $x \longrightarrow \frac{x}{4} - 3$ is $x \longrightarrow 4(x + 3)$.
   c $\frac{x}{7} + 2 \longleftarrow$ add 2 $\longleftarrow$ divide by 7 $\longleftarrow x$
      The inverse of $x \longrightarrow 7(x + 2)$ is $x \longrightarrow \frac{x}{7} + 2$.
   d $5x - 6 \longleftarrow$ subtract 6 $\longleftarrow$ multiply by 5 $\longleftarrow x$
      The inverse of $x \longrightarrow \frac{x + 6}{5}$ is $x \longrightarrow 5x - 6$.

C   **a** $x \longrightarrow \frac{x}{4}$ **b** $x \longrightarrow x + 8$ **c** $x \longrightarrow \frac{x + 2}{3}$
   d $x \longrightarrow \frac{x}{2} - 1$ **e** $x \longrightarrow 2(x + 2)$ **f** $x \longrightarrow 6 - x$

51 Graphing functions   page 53

 A **a** **i**

x	0	4	⁻4
y	3	5	1

ii

x	0	4	⁻4
y	⁻1	1	⁻3

b

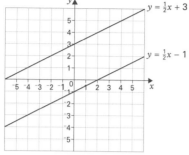

$y = \frac{1}{2}x + 3$

$y = \frac{1}{2}x - 1$

c No, because $9 \neq \frac{1}{2}(10) + 3$

d ⁻6

e They have the same gradient

B **a** **i** C   **ii** E   **iii** A   **iv** B   **v** D

b Yes, because $2 = {}^-2(2) + 6$

c 3

C **a** A: $y = 3x - 1$ B: $y = 4x + 2$ C: $y = 6x + 4$

D: $y = 3x + 2$

b A and D

c $y = 2x - 1$, $9y = 3x - 9$, $\frac{y+1}{3} = x$

52 Gradient of a straight line   page 54

A l_1 gradient $= \frac{4}{8} = \frac{1}{2}$ l_2 gradient $= \frac{^-6}{2} = {}^-3$

l_3 gradient $= \frac{1}{3}$

l_4 gradient $= \frac{^-1}{1}$ or $\frac{^-2}{2}$ or $\frac{^-3}{3}$ etc. $= {}^-1$

B **a** C, D, E   **b** A, B   **c** 0

d **i** C   **ii** A   **iii** F   **iv** D   **v** E   **vi** B

C Line R: gradient $^-\frac{1}{2}$, equation $y = {}^-\frac{1}{2}x - 2$

Line S: gradient $^-4$, equation $y = {}^-4x + 5$

Line T: gradient $\frac{2}{3}$, equation $y = \frac{2}{3}x + 5$

53 Distance–time graphs   page 55

 A **i** D   **ii** A   **iii** B   **iv** C

 B **a** 11 am   **b** 10 km

c **i** 30 minutes   **ii** 4 km

d 15 minutes

e From 19 km to 26 km, because they were travelling down hill.

f

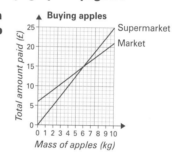

Cycling journey

54 Real-life graphs   page 56

 A **a**

b

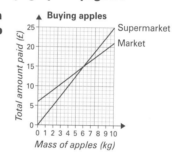

Buying apples

Supermarket

Market

c **i** Supermarket   **ii** Market

d 6 kg

B **a** **i** Graph 2, because depth will increase steadily with time.

ii Graph 3, because depth will steadily increase but once it reaches the thin part of the container it will steadily increase more quickly.

iii Graph 1, because the depth will increase at a faster and faster rate with time.

b **i**   **ii**

Answers – Shape, Space and Measures

55 Conventions, definitions and derived properties page 58

a D   **b** DP   **c** C   **d** C
e D   **f** DP   **g** D   **h** C
i DP

a Convention: We use two dashes on sides of a shape to show they are equal.
Definition: A kite is a quadrilateral with two pairs of adjacent equal sides.
Derived property: The diagonals of a kite meet at right angles.

b Convention: If no direction is given for a rotation we rotate anticlockwise.
Definition: A rectangle is a quadrilateral with two pairs of opposite sides equal and all angles equal to 90°.
Derived property: The diagonals of a rectangle bisect each other or the diagonals of a rectangle are equal.

56 Finding angles page 59

a $a = 55°$ **b** $b = 118°$
c $c = 50°, d = 45°$ **d** $e = 105°, f = 75°$
e $g = 65°, h = 65°, i = 65°$
f $j = 58°, k = 104°$

B
a $a = 50°$ Corresponding angles on parallel lines are equal
$a + 80° + x = 180°$ Angles in a △ add to 180°
$50° + 80° + x = 180°$ Substituting $a = 50°$
$130° + x = 180°$
$x = 180 - 130°$ Subtracting 130° from both sides
$x = 50°$

b label angles a and b

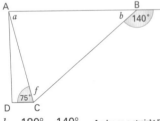

$b = 180° - 140°$ Angles on a straight line add to 180°
$= 40°$
$a = 75°$ Alternate angles on parallel lines are equal
$f + b + a = 180°$ Angles in a △ add to 180°
$f + 40° + 75° = 180°$ Substituting $b = 40°$ and $a = 75°$
$f + 115° = 180°$
$f = 180° - 115°$ Subtracting 115° from both sides
$f = 65°$

57 More finding angles page 60

a label angles a, b, c and d

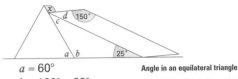

$a = 60°$ Angle in an equilateral triangle
$b = 180° - 60°$ Angles on a straight line add to 180°
$= 120°$
$c = 180° - 25° - 120°$ Angles in a △ add to 180°
$d = 35°$
$2d = 360 - 2 \times 150°$ Angles in parallelogram add to 360°
$d = 30°$
$x = 180° - 30° - 35°$ Angles on a straight line add to 180°
$= 115°$

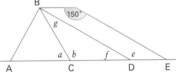

b $a = 60°$ and $b = 120°$ from part **a** label e, f and g
$e = 150°$ Parallelograms have rotation symmetry of order 2
$f = 180° - 150°$ Angles on a straight line add to 180°
$= 30°$
$g = 180° - 120° - 30°$ Angles in a △ add to 180°
$= 30°$
$f = g$ so triangle BCD is isosceles.

B
Label s
$s = \frac{360}{8}$ Angles at a point add to 360°
$= 45°$
Each triangle is isosceles so
$s + t + t = 180°$
$45 + 2t = 180°$
$2t = 135°$
$t = 67·5°$

C
a $5x + 15 + 3x - 3 = 180°$
$x = 21°$
b $3x + 4 = 4x - 8$
$x = 12°$

D
a Cut angle 92° with a parallel line. Label these new angles x and y.
$x = 41°$ Alternate angles are equal
$y = 92° - 41°$
$= 51°$
$a = 51°$ Alternate angles are equal

b Cut angle b with a parallel line. Label angles x, y and z.
$x = 76°$ Alternate angles on parallel lines are equal
$z = 180° - 76°$ Angles on a straight line add to 180°
$= 104°$
$y = 38°$ Alternate angles on parallel lines are equal
$b = 104° + 38°$
$= 142°$

58 Interior and exterior angles of a polygon   page 61

A   **a** 100°   **b** 36°   **c** 63°   **d** 71°

B   The angles in a triangle add up to 180°.
A quadrilateral can be split into two triangles, so the angles in a quadrilateral add up to 2 × 180° = 360°.

C

Shape	Number of sides	Sum of interior angles	Size of one interior angle	Sum of exterior angles	Size of one exterior angle
Square	4	360°	90°	360°	90°
Regular hexagon	6	720°	120°	360°	60°
Regular octagon	8	1080°	135°	360°	45°
Regular 15–sided polygon	15	2340°	156°	360°	24°

D   **a** 112°   **b** 70°

59 Triangles and quadrilaterals   page 62

A   **a** F   **b** T   **c** T   **d** F   **e** T

B   **a** $a = 180° - 125°$ Angles on a straight line add to 180°
      $= 55°$
      $b = 55°$ A rhombus is symmetrical about its diagonal
   b $c = 62°$ Corresponding angles on parallel lines are equal
      $d = 180° - 2 \times 62°$ Base angles in an isosceles △ are equal
      $= 56°$ Angles in a △ add to 180°

C   Because all squares have two pairs of equal sides and 4 right angles so they satisfy the definition of a rectangle.

D   **a**

parallelogram   equilateral triangle   hexagon

   b

arrowhead   parallelogram

E   **a** 10 cm   **b** 10·4 m   **c** 27·5 mm   **d** 5·8 m

60 Tessellations   page 63

A   *Possible answers are:*
   a
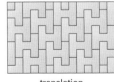
translation
   b

translation
   c

rotation and translation

rotation and translation

B   **a i**

Regular polygons which tessellate	Size of one internal angle
Equilateral triangle	60°
Square	90°
Hexagon	120°

   ii Tessellating regular polygons have internal angles which are factors of 360°. This means they will meet at a point to give 360°.

   b i

(squares and equilateral triangles)

   ii
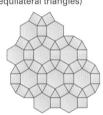
(squares and regular octagons)

   iii
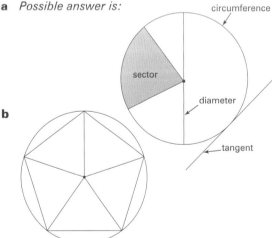
(regular hexagons, squares and equilateral triangles)

61 Circles   page 64

A   **a** Centre   **b** Radius   **c** Semicircle
   d Circumference   **e** Tangent   **f** Arc
   g Segment   **h** Diameter   **i** Chord
   j Sector

B   **a** Arc   **b** Chord
   c Segment   **d** Tangent

C   **a** *Possible answer is:*
circumference
sector
diameter
tangent

   b

62 Constructions   page 65

A   **b** 97°

B   **b** About 7·3 cm

C   **b** About 68·5 m

D   **a** No, because angles in a triangle add to 180°, not 170°.
   b Yes

63 More constructions   page 66

B **b i** About 6·2 m   **ii** About 57°

C **b** 8 m   **c** About 37°

64 Locus   page 67

A **a i** C   **ii** A   **iii** B
       b i B   **ii** C   **iii** A

B

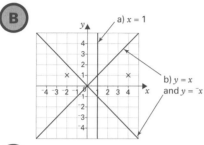

a) x = 1

b) y = x and y = ⁻x

C **a**

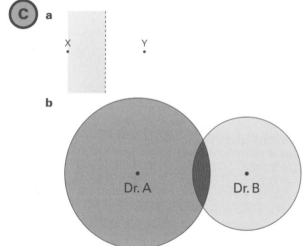

b

Dr. A   Dr. B

65 3-D shapes   page 68

A **i** Side view   **ii** Plan view   **iii** Front view

B **a i**   **ii**   **iii**

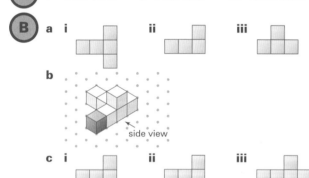

b

side view

c i   **ii**   **iii**

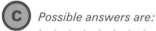

C *Possible answers are:*

D **a**   **b**   **c**

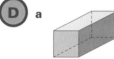

66 Cross-sections   page 69

A **a** Triangle
       b She would have got a larger triangle.
       c She would have got a smaller triangle.

B **a i** Rectangle
        ii He would have got a smaller rectangle.
       b Triangle

C **a** Circle   **b** Rectangle
       c Oval or ellipse

D **a**

b *Possible answers are:*
   i   **ii**

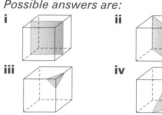

   iii   **iv**

67 Congruence and transformations   page 70

A **a** DEF and VWX
       b i DE and WX   **ii** EF and VW
       c i F and V   **ii** E and W

B **a i** 8 cm   **ii** 20°   **iii** 80°
       b i Angles in a triangle add to 180°, so ∠SQP
              must be 80° and if a triangle has two equal
              angles it must be isosceles.
         ii 80°
        iii 8 cm

C **a i** Transformation 1
       b A is congruent to A′
         ii Transformation 7
              A is congruent to A′
        iii Transformation 4
              A is congruent to A′

68 More congruence and transformations   page 71

A **a i**

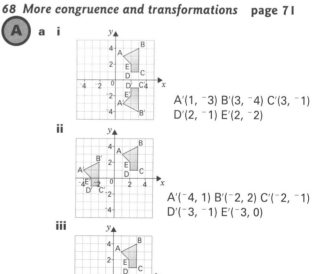

A′(1, ⁻3) B′(3, ⁻4) C′(3, ⁻1)
D′(2, ⁻1) E′(2, ⁻2)

   ii

A′(⁻4, 1) B′(⁻2, 2) C′(⁻2, ⁻1)
D′(⁻3, ⁻1) E′(⁻3, 0)

   iii

A′(⁻1, ⁻3) B′(⁻3, ⁻4) C′(⁻3, ⁻1)
D′(⁻2, ⁻1) E′(⁻2, ⁻2)

b Yes
c **i** Yes　**ii** Yes　**iii** Yes

B **a**

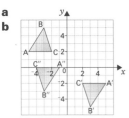

b **i** (⁻3, 0), (0, 4), (3, 1)
　ii (4, ⁻2), (1, 2), (⁻2, ⁻1)
　iii (⁻2, 2), (2, ⁻1), (⁻1, ⁻4)
c All of them

C **a** **i** Three of CD, FG, GH, JK, KL, NO, OP
　ii Three of AP, AB, DE, EF, HI, IJ, LM
b **i** Three of ∠ABC, ∠CDE, ∠EFG, ∠GHI, ∠IJK, ∠KLM, ∠MNO
　ii Three of ∠LKQ, ∠NOQ, ∠POQ, ∠BCQ, ∠DCQ, ∠FGQ, ∠HGQ

69 Combinations of transformations　page 72

A **a** Kite or arrowhead
b Rectangle
c Rhombus

B B

C **a**
b

c D
d Rotation of 180° about (⁻3, 1)

D *Possible answers are:*
a **1** Reflection in *x*-axis
　2 Translation 2 units left
b **1** Rotation 180° about the origin
　2 Translation 3 units right and 1 unit up
c **1** Rotation 270° about the origin
　2 Reflection in the line *y* = 2

70 Symmetry　page 73

A **a** 3 lines of symmetry
　　rotation symmetry of order 3
b 0 lines of symmetry
　　rotation symmetry of order 5
c 4 lines of symmetry
　　rotation symmetry of order 4

B *a* = 130°, *b* = 40°, *c* = 140°

C **a**

b

71 Enlargement　page 74

A

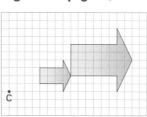

B **a** 3　**b** 2·5

C **a** *A* = 101°, *x* = 8 cm, *y* = 7 cm
b *A* = 132°, *x* = 15 cm, *y* = 6 cm

D **a** 15 m　**b** 75 cm

72 Making scale drawings　page 75

A **a** 1 : 5　**b** **i** 250 000　**ii** 1 : 100 000
c **i** 11 000　**ii** 1 : 2000

B The room will be 3·5 cm by 2·2 cm.
The bed will be 2 cm long and 1·1 cm wide.
The desk will be 1·4 cm long and 0·6 cm wide.
The wardrobe will be 0·9 cm long and 0·7 cm wide.
The drawers will be 0·8 cm long and 0·8 cm wide.

C **a** **i** 1 m　**ii** 50 cm　**iii** 5 cm
b **i** 18 cm　**ii** 9 cm　**iii** 3·6 cm
c **i** 150 cm　**ii** 15 cm　**iii** 30 cm

73 Interpreting scale drawings　page 76

A **a** About 0·9 m
b About 3·6 m
c About 5·4 m

B **a** 2 cm　**b** 2 m　**c** 2·2 m
d 2·6 m　**e** 65°　**f** 11·4 m

C **a** **i** 4 cm　**ii** 20 km
b 17·5 km
c 7000 m
d Paradise Beach

74 Metric conversions, including area, volume and capacity　page 77

A ABOUT ONE BILLION SNAILS ARE SERVED IN RESTAURANTS EACH YEAR

B **a** **i** 187　**ii** 18 700　**iii** 0·0187
b **i** 1125　**ii** 1 125 000
　iii 1125　**iv** 1·125

75 Working with measures　page 78

A **a** 50p/m. There are about 3 feet in a metre, so 20p/foot is about 60p/m.
b 2p/cm　**c** 80p/km
d £2/pound　**e** 175/oz
f £1·50/ℓ　**g** £2·80/gallon

B **a** 3 km　**b** 080°　**c** 260°

C
a 80 km/hr
b 60 km/hr
c Average speed = $\frac{\text{total distance}}{\text{total time}} = \frac{50 \text{ km}}{45 \text{ min}} = \frac{50 \text{ km}}{0\cdot75 \text{ hr}}$
= 66.7 km/hr (1 d.p.)

76 Perimeter and area   page 79

A Possible answers are:

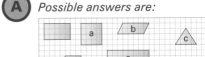

B
a $A = 135$ cm^2, $P = 54$ cm
b $A = 76.245$ m^2, $P = 38\cdot9$ m
c $A = 173\cdot04$ mm^2, $P = 53$ mm

C
a Order is C, A, B
b i 5   ii 125 m

77 Circumference   page 80

A
a 62·8 mm   b 43·96 cm   c 28·26 m
d 40·84 cm   e 30·16 m   f 102·42 mm

The following answers have been given for using π on the calculator.

B
a 5·65 m (2 d.p.)
b 50·93 cm
c 125·7 cm (1 d.p.) or 125·66 cm (2 d.p.)

C
a 18·85 m
b 20

D
a 35·7 m (1 d.p.)
b 42·8 (1 d.p.)

78 Area of a circle   page 81

Answers have been given for using π on the calculator.

A
a 153·9 m^2 b 260·2 cm^2
c 530·9 mm^2 d 120·8 mm^2

B
a 72·4 m^2 b 120·8 m^2 c 79·8 m^2

C
a i Choice 2   ii 78·5 cm^2
b 235·6 cm^2

D
a i 79 m^2 ii 5 m
b i 166·8 mm^2 (1 d.p.)
ii 131·4 cm^2 (1 d.p.)

79 Surface area and volume of a prism   page 82

A
a $V = 308$ mm^3, $SA = 298$ mm^2
b $V = 15$ m^3, $SA = 42$ m^2
c $V = 8784$ mm^3

B
a 120 cm^3 b 6 cm

C
a 20·52 m^2 b 82·08 m^2 c 95·04 m^2

Answers – Handling Data

 80 Planning a survey   page 84

Answers will vary. *Possible answers are:*

 a **A 1.** Males can swim faster than females.
   2. The greater the time spent practising, the faster the swimmer.

   Conjecture 2
   B i Average time spent practising each week.
      Time taken to swim 50 m.
   ii Survey – ask students how long they spend practising.
      Experiment – measure the time taken to swim 50 m.
   iii Primary
   C i At least 100 people, because a large sample size is needed to be representative of all students.
   ii Practice time to nearest half hour.
      50 m swimming time to the nearest second.

b **A 1.** The higher the wattage the less time the light bulb lasts for.
   2. Bulbs greater than 75 w last only half as long as those less than 40 w.

   Conjecture 1
   B i Wattage of bulbs.
      Time each bulb lasts for.
   ii Experiment or manufacturer's information.
   iii Experiment = primary
      Manufacturer's information = secondary
   C i 100–200, because a large sample size is needed to be accurate.
   ii Wattage to the nearest 5 watts.
      Time to the nearest hour.

81 Collecting data   page 85

 Possible answers are:
a **1.** The sample size is too small.
   2. Dougal's dad's work might not be representative of all people.
b **1.** Carey's sign might attract a biased proportion of people who **do** own a mobile phone.
   2. People at a school may not be representative of all people.

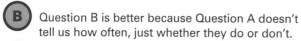 Question B is better because Question A doesn't tell us how often, just whether they do or don't.

 Possible answers are:
a Write the names of each of your school subjects below.
   Beside each subject, write one of these numbers:
   1. I really enjoy it.
   2. I quite enjoy it.
   3. It's OK.
   4. I don't like it much.
   5. I strongly dislike it.

b I get to school on time.
   Always ☐   Usually ☐   Sometimes ☐
   Not often ☐   Never ☐
c I complete my homework on time.
   Always ☐   Usually ☐   Sometimes ☐
   Not often ☐   Never ☐

82 Collecting data   page 86

 a Callum's, because it is time consuming to write the whole word each time, and because the data will still need to be grouped later.
b Dale's, because it is a fast way to record data, and the data is grouped immediately.

 Possible answers are:
a **1.** Names are not needed.
   2. Data could be grouped.
b

Length of foot (cm)	Tally	Frequency
$16 < l \leqslant 18$		
$18 < l \leqslant 20$		
$20 < l \leqslant 22$		
$22 < l \leqslant 24$		
$24 < l \leqslant 26$		
$26 < l \leqslant 28$		
$28 < l \leqslant 30$		

 Possible answers are:
a

Height (m)	Tally	Frequency
$1 \cdot 3 < h \leqslant 1 \cdot 4$		
$1 \cdot 4 < h \leqslant 1 \cdot 5$		
$1 \cdot 5 < h \leqslant 1 \cdot 6$		
$1 \cdot 6 < h \leqslant 1 \cdot 7$		
$1 \cdot 7 < h \leqslant 1 \cdot 8$		
$1 \cdot 8 < h \leqslant 1 \cdot 9$		

b

Number of hours sleep	Tally	Frequency
$h < 5$		
$5 \leqslant h < 6$		
$6 \leqslant h < 7$		
$7 \leqslant h < 8$		
$8 \leqslant h < 9$		
$\geqslant 9$		

83 Two-way tables   page 87

 a 26   **b** 92   **c** 30
d Upper, 106
e More junior school pupils were born in Europe, and less were born in America. About the same numbers of junior and middle school pupils were born in Africa, Asia and the Pacific.

 a Year 9
b Pupils were able to run more laps as they got older.
   In general, more boys ran 6 or 7 laps than girls. Less boys ran 3 or 4 laps than girls. About the same proportions of boys and girls ran 5 laps. Overall, the boys ran more laps than the girls.

	Polarfleece	Wool
Drying time		
Stretch		
Durability		
Insulation		

84 Mode, median, mean   page 88

 A
a Mean 502 g, median 503·5 g, mode 513 g
b Yes, because the data is spread evenly around the mean.
c No, because there is only one value above 513 g.
d 510–519 g

 B
a Mean 35·7 (1 d.p.), median 48, mode 0
b Median, because the mean is lowered by the two zero values and 5 out of 7 values are higher than the mode. The median is where most values are clustered around.

 C **a** B   **b** D

 D **a** 4, 4, 10   **b** 7, 5, 12
c *Possible answers are:* 2, 5, 8 or 1, 7, 7 or 3, 3, 9

85 Displaying data on bar charts and frequency diagrams   page 89

 A
a Discrete
b Compound bar chart, because this can display wins, draws and losses of all four teams on one graph.
c

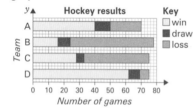

d B
e Team A had more wins and fewer losses than team B.
Both teams had similar numbers of draws.

 B
a Because the data is continuous.
b

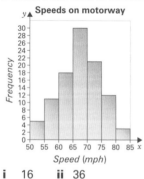

c **i** 16   **ii** 36
d 65–

86 Displaying data on line graphs and pie charts   page 90

 A **a**

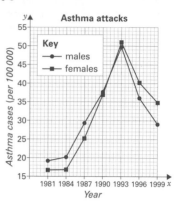

b 1993   **c** 1990–1993
d From 1981 to 1990 males recorded more asthma cases than females.
From 1993 to 1999 the opposite was true.
e Trends over time are best displayed on a line graph.

 B **a** Wood 54°, Gas 151°, Electricity 88°, Other 67°

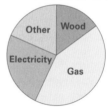

b 135

87 Scatter graphs   page 91

 A
a A positive correlation
b The point (22, 125) is not close to the trend of the other points.
c Somewhere between 108 cm and 118 cm

 B **a**

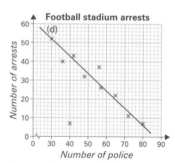

b A negative correlation
c (40, 7)   **e** About 70

88 Interpreting graphs   page 92

 A
a 27   **b** Not possible to tell
c Class A, because there are a greater proportion of students with heights > 160 cm.

 B
a **i** About 32%   **ii** About 9%
b About 800
c Generally, the female runners were younger than the male runners.

89 More interpreting graphs   page 93

a $250 \leqslant d < 300$ **b** $150 \leqslant d < 200$
c Yes, because only three people jumped further than 350 cm. 22 people jumped in the interval $300 \leqslant d < 350$, but it is possible that all except Jack jumped < 320 cm.
d Yes, because the median is in the interval $200 \leqslant d < 250$, so Sophie's jump of 251 cm is above the median.

a Vegetables
b It went up from about 500 g to about 700 g.
c It went down from about 1300 g to less than 700 g.
d 1979–1984

a No. The percentages may have all been rounded up resulting in a rounded total of 101%.
b False. Sea World has 29% of 14 = 4 hooded seals and Ocean Park has 19% of 37 = 7 hooded seals.
c Ocean Park, 6 more.

90 Misleading graphs   page 94

a Graph 1   **b** No
c Graph 1, because it gives the impression of rapidly rising sales.

a The vertical scale does not start at zero, so it makes the price drop look larger than it really is.
b There is no vertical scale, so the heights of the bars cannot be compared.
c Making the 'bars' 3-D makes the increase look greater than it really is.
d Only 18 people have been surveyed. This is a very small sample.

91 Comparing data   page 95

a

Jane

Time (secs)	Tally	Frequency
300–309	II	2
310–319	III	3
320–329	II	2
330–339	III	3
340–349	II	2
350–359	II	2

Hannah

Time (secs)	Tally	Frequency
300–309	I	1
310–319	III	3
320–329	III	3
330–339	HM II	7
340–349		
350–359		

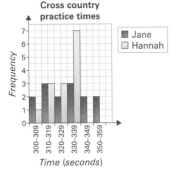

b Jane's graph shape is more spread out. Hannah's graph shape is less spread and has lots of times in the 330–339 sec interval.

c   **Jane**   Mean 328·1 (1 d.p.).
Median 327   Range 58
  Hannah   Mean 325·6 (1 d.p.).
Median 327   Range 30
d *Possible answer is:*
Hannah because her mean time is slightly faster and her times are more consistent, as shown by the smaller range.

93 Language of probability   page 97

a Disagree, because there is a 50% chance of getting a girl each time she has a baby.
b *Possible answer is:*
Disagree, because more males than females buy football tickets.
c Disagree, because his ticket has the same chance of winning as every other ticket.

Box A

a Screen 2   **b** Screen 5

94 Mutually exclusive events   page 98

a Mutually exclusive
b Mutually exclusive
c Not mutually exclusive
d Mutually exclusive
e Not mutually exclusive
f Mutually exclusive

a Not mutually exclusive
b Not mutually exclusive
c Mutually exclusive

a A heart, a number less than 4
b A circle, an even number
c A prime number, a number greater than 8

95 Calculating probabilities of mutually exclusive events   page 99

1 i $\frac{3}{20}$ ii $\frac{1}{2}$ iii $\frac{4}{5}$
2 **a** $\frac{5}{16}$
  b i 5   ii 4   iii 2   iv 5
3 7

B

a i 0·15   ii 0·35
b 300

C

a $\frac{2}{5}$ **b** Centaur   **c** $\frac{14}{15}$ **d** 60

96 Listing mutually exclusive outcomes   page 100

A

a

Spinner 1	○	○	○	△	△	△	□	□	□
Spinner 2	○	△	□	○	△	□	○	△	□

b ○○, ○△, ○□, △○, △△, △□, □○, □△, □□

B

a HH, HT, TH, TT
b HHH, HHT, HTH, HTT, TTT, TTH, THT, THH

a

+	1	2	3	4	5	6
1	2	3	4	5	6	7
2	3	4	5	6	7	8
3	4	5	6	7	8	9
4	5	6	7	8	9	10
5	6	7	8	9	10	11
6	7	8	9	10	11	12

b 7, because there are more ways of throwing a 7 than any other number.

D **a** **i** ABC, ABD, ABE, BCD, BCE, CDE
ii 6
b **i** EABC, EABD, EBCD
ii 4

97 Calculating probability by listing all the mutually exclusive outcomes page 101

A **a** sun/sun, sun/moon, moon/sun, moon/moon
b **i** $\frac{1}{4}$ **ii** $\frac{1}{2}$ **iii** $\frac{3}{4}$

B **a**

	5p	10p	20p	50p
5p	10p	15p	25p	55p
10p	15p	20p	30p	60p
20p	25p	30p	40p	70p
50p	55p	60p	70p	£1

b **i** $\frac{1}{16}$ **ii** $\frac{9}{16}$ **iii** $\frac{1}{2}$

C **a** GRC, GRW, GBC, GBW, SRC, SRW, SBC, SBW
b **i** $\frac{1}{8}$ **ii** $\frac{1}{4}$ **iii** $\frac{1}{4}$ **iv** $\frac{3}{8}$

98 Estimating probabilities from relative frequency page 102

A **a** 60
b **i** $\frac{1}{3}$ **ii** $\frac{1}{4}$
c $\frac{13}{60}$
d 40

B **a** **i** $\frac{1}{5}$ **ii** $\frac{3}{40}$
b **i** $\frac{47}{100}$ or about $\frac{1}{2}$
ii survey more people
c About 80

C **a** 200 **b** 20
c Sven's because he played more games.

99 Comparing experimental and theoretical probability page 103

A **a** **i** $\frac{1}{4}$ **ii** 5
iii A possible answer is: Yes, because this is a little greater than you would expect even with random variation.
b No, because you would expect to win about 170 times, and with random variation 166 is quite close to 170.

B **a**

+	2	3	4	6
1	3	4	5	7
2	4	5	6	8
3	5	6	7	9
5	7	8	9	11

b Bernard, because the sample space has more odd numbers than even, square or prime numbers.
c Bernard, Finbar, Hilary, Kelly

Answers – Practice Papers

Paper 1 **page 105**

1 **a** $\frac{1}{3}$ **b** $\frac{1}{6}$ **c** $\frac{1}{6}$ **d** $\frac{7}{24}$

2 **a** *One possible answer is:*

b Parallel

3 **a** June, 19
b December, about $5\frac{1}{2}$

4 **a** $\frac{1}{40}$ **b** $\frac{17}{40}$ **c** $\frac{23}{40}$

5 **a** £40·50 **b** £38·25

6 **a** **i** £3·80 **ii** £4·20
b 40p
c

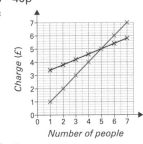

d 5

7 **a**

8	7	12
13	9	5
6	11	10

b $x = 14$, $y = 7$, $z = 2$

8 **a** **i**

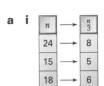

ii

n	→	$n - 6$
9	→	3
23	→	17
17	→	11

b *Possible answers are:*
i

n	→	$3n$
4	→	12

ii

n	→	$n + 8$
4	→	12

9 **a** All the same **b** Brick A **c** 8
d

	Brick 1	Brick 2	Brick 3	Brick 4	Brick 5
Length	10	5	30	15	5
Width	3	6	1	2	3
Height	1	1	1	1	2

10

A	4	6	⁻3
B	7	⁻2	⁻4
Sum of A and B	11	4	⁻7
Product of A and B	28	⁻12	12

11 $\frac{1}{2}$

12 **a** **i** $y = x + 2$ **ii** $f = \frac{e}{3}$ **iii** $r = 6q - 5$
b $n = \frac{p}{4} + 3$

13 100 km/h

Paper 2 **page 109**

1

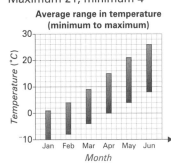

2 **a** Maximum 21, minimum 4
b

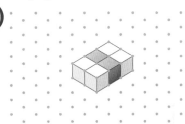

3 200

4 **a** A prime number has no factors except itself and 1. 35 also has factors 5 and 7.
b All of the numbers in column Y are multiples of 6. They all have 2 and 3 and 1 as factors.
c All of the numbers in column X are multiples of 3. They all have 3 as a factor.

5 Yes, because the rule shows that both males were 186 cm tall.

6 **a** £708·75 **b** 5·1% (1 d.p.)

7   **a** $y = 8$ cm   **b** $w = 3$ cm

8   **a**   *Possible answer is:*

(x, y)	$(^-1, 5)$	$(0, 4)$	$(1, 3)$
$x + y$	4	4	4

   b $x + y = 4$

   c

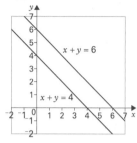

9   **a**   Maximum 45, minimum $^-15$

   b   14

   c   *Possible answers are:*

Number of jumps attempted	7	15	11
Number of successful jumps	7	9	8
Number of bars knocked down	0	6	3

10   **a**   25·1 cm (1 d.p.)

    b   107 times if we use the unrounded circumference from **a**.

11   **a**   10 mℓ **b**   50 mℓ

12   18 points

13   **a** $\frac{3\,\ell}{\text{tea}}$ $\frac{6\,\ell}{\text{ginger ale}}$ $\frac{13\cdot5\,\ell}{\text{juice}}$

    b $\frac{2\cdot4\,\ell}{\text{tea}}$ $\frac{4\cdot9\,\ell}{\text{ginger ale}}$ $\frac{11\,\ell}{\text{juice}}$